Concrete Durability

Luis Emilio Rendon Diaz Miron • Dessi A. Koleva
Editors

Concrete Durability

Cementitious Materials
and Reinforced Concrete Properties,
Behavior and Corrosion Resistance

 Springer

Editors
Luis Emilio Rendon Diaz Miron
Universidad Internacional (UNINTER)
Cuernavaca, Morelos, Mexico

Dessi A. Koleva
Delft University of Technology
Faculty of Civil Engineering
  and Geosciences
Section Materials and Environment
Delft, Zuid-Holland, The Netherlands

ISBN 978-3-319-85668-1   ISBN 978-3-319-55463-1   (eBook)
DOI 10.1007/978-3-319-55463-1

Printed on acid-free paper

This Springer imprint is published by Springer Nature
The registered company is Springer International Publishing AG
The registered company address is: Gewerbestrasse 11, 6330 Cham, Switzerland

Preface

In the introduction of his book *The Substance of Civilization: Materials and Human History from the Stone Age to the Age of Silicon*, Stephen L. Sass indicates: "Materials not only affect the destinies of nations but define the periods within which they rise and fall. Materials and the story of human civilization are intertwined, as the naming of eras after materials – the Stone age, the Bronze age, the Iron age – remind us."

Indeed, we can say that we are in the Concrete age, considering how concrete is the most widely known man-made material in the world, with annual consumption estimated at between 21 and 31 billion tons. Next to the fact that worldwide infrastructure is mainly composed of concrete and reinforced concrete, this man-made material is omnipresent in the everyday life – people use it, rely on it, and live on or in it. Concrete is a moldable composite material in which natural gravel stones of well-graded sizes are bound together by a cement matrix, which provides concrete strength. Although this principle of building materials fabrication can be traced back to at least the Roman Empire, it was later neglected for centuries. Concrete regained importance in the 1800s, when methods for large-scale production of modern cement were developed; since then, its use has expanded tremendously.

Concrete is a durable and sustainable material, widely used for both buildings and infrastructure. When properly designed and constructed, it can resist severe storms and earthquakes, as well as aggressive environments; nonetheless, concrete is a brittle material. Reinforced with steel bars (reinforced concrete) or with prestressed steel tendons (prestressed concrete), its applicability greatly expands. However, reinforced concrete is also prone to various degradation mechanisms, primarily corrosion of the steel reinforcement, which impacts durability and service life. This leads to a strong financial, environmental, and safety concerns. This creates an increasing scientific and engineering focus to extend the service life of reinforced concrete structures. Novel solutions for new cement blends, composite coatings, nanomaterials, self-healing approaches, etc., are more recently studied solutions and show great potential. Nevertheless, these methods often target the quality of the cement-based material only, or solely enhanced steel corrosion resistance, rather than considering the overall complexity of "reinforced concrete" as a composite system. To this end, novel, feasible, and sustainable solutions, originating from entirely different scientific backgrounds, are of scientific interest and are being explored for practical applications.

The importance of reinforced concrete as a building material is due to its structural, physical, and chemical properties, and its cost-efficiency and effectiveness. In one's hands, concrete is a sentient entity, capable of creating amazing forms while providing resistant structures. It is a material that has been able to endure modern demands. To craft good concrete, one needs to study hard, get out to the field, get dirt in one's shoes, and treat it with affection and respect; this is the only language that this unique material understands.

This uniqueness of concrete may make it appear to be an awkward material in light of today's scientific advances. In fact, all concrete may seem the same. Certainly, the basic product has remained unchanged since its invention. Nonetheless, concrete properties depend largely on the exploration conditions (e.g., environment) and the quantity and quality of its components, including Portland cement. The selection, use, and constituents of components are important to design appropriately, and as economically as possible, the desired characteristics of any particular type of concrete.

This book disseminates new knowledge on concrete durability for the field of civil engineering. It collects recent studies on cementitious materials, including reinforced concrete properties, behavior, and corrosion resistance. Biodegradation of concrete structures is discussed, as well as the utilization of wastes to control such degradation mechanisms. The major durability-related challenge for reinforced concrete structures, i.e., chloride-induced steel corrosion, is presented in view of recent studies on sensors and sensor technology for early corrosion detection, but also with regard to novel methods for corrosion control, such as the application of hybrid nanomaterials. Furthermore, the rather seldom reported in the present state of the art, stray current-induced corrosion is discussed with regard to both steel reinforcement and bulk matrix properties and performance. This book aims at contributing to the present knowledge on the subject, raising awareness on the complex and challenging aspects of materials' behavior within service life of reinforced concrete structures. It targets dissemination of fundamentally substantiated mechanisms and verification of innovative applications, in view of bringing confidence for industrial utilization of novel solutions and practices in civil engineering.

The research contributions in every chapter of this book highlight new technologies to reverse the trend in the weathering, decaying concrete infrastructure, and will be of interest to academics, engineers, and professionals involved in concrete and concrete infrastructure.

Cuernavaca, Morelos, Mexico   Luis Emilio Rendon Diaz Miron
Delft, Zuid-Holland, The Netherlands   Dessi A. Koleva

Contents

Chapter 1
The Effect of Microorganisms on Concrete Weathering

Luis Emilio Rendon Diaz Miron and Montserrat Rendon Lara

Abstract Concrete structures exposed to aggressive aqueous media (waste water, soft water, fresh water, ground water, sea water, agricultural or agro-industrial environments), due to their porous nature, are susceptible to a variety of degradation processes resulting from the ingress and/or presence of water. In addition to chemical and physical degradation processes, the presence of water contributes to undesirable changes in the material properties resulting from the activities of living organisms, i.e., biodeterioration. Since microorganisms are ubiquitous in almost every habitat and possess an amazingly diversified metabolic versatility, their presence on building materials is quite normal often, they can infer deterioration that can be detrimental (loss of alkalinity, erosion, spalling of the concrete skin, corrosion of rebars, loss of water- or air tightness, etc.). The deleterious effect of microorganisms, mainly bacteria and fungi, on the cementitious matrix has been found to be linked, on the one hand, with the production of aggressive metabolites (acids, CO_2, sulfur compounds, etc.) but also, on the other hand, with some specific, physical and chemical effects of the microorganisms themselves through the formation of biofilm on the surface. Moreover, the intrinsic properties of the cementitious matrix (porosity, roughness, mineralogical and/or chemical composition) can also influence the biofilm characteristics, but these phenomena have not been understood thoroughly as of yet.

These deteriorations lead to a significant increase in the cost of repairing structures and to loss of production income, but may also lead to pollution issues resulting, for example, from waste water leakage to the environment. Also, building facades, and notably concrete external walls, can be affected by biological stains, which alter aesthetical quality of the construction, sometimes very quickly, and lead to significant cleaning costs. Microorganisms, mainly algae, responsible for these alterations have been quite well identified. Research is now rather focused on determining colonization mechanisms, and notably influencing material-related factors, and on development of preventive or curative, and preferentially environmentally

L.E. Rendon Diaz Miron (✉)
Universidad Internacional (UNINTER), Cuernavaca, Morelos, Mexico
e-mail: luisrendon@uninter.edu.mx

M.R. Lara
Centre of Arts of the State of Morelos, Cuernavaca, Morelos, Mexico

© Springer International Publishing AG 2017   1
L.E. Rendon Diaz Miron, D.A. Koleva (eds.), *Concrete Durability*,
DOI 10.1007/978-3-319-55463-1_1

friendly, solutions to protect external walls. However, up to now, no clear results about the efficiency of these various protection solutions are available.

Keywords Concrete Weathering • Microorganisms • Deterioration

1.1 Introduction

Concrete is probably one of the oldest engineering and structural materials. There are many types of concrete available, created by varying the proportions of the main ingredients. In this way or by substitution for the cementitious and aggregate phases, the finished product can be tailored to its application with varying strength, density, or chemical and thermal resistance properties. At the same time, concrete can be damaged by many processes, such as the expansion of corrosion products of the steel reinforcement bars, freezing of trapped water, fire or radiant heat, aggregate expansion, sea water effects, bacterial corrosion, leaching, erosion by fast-flowing water, physical damage and chemical damage. Bacterial corrosion also known as microbiologically induced corrosion of concrete (MICC) or biodeterioration of concrete is one of the most serious problems, most of these concrete failures occur in sewerage works. Furthermore, most of the measurements to prevent concrete failures are design to be used in hydraulic infrastructure to avoid repairs and replacements. Not only is the replacement of sewer pipes very expensive, but also sewer-pipe failure causes leaking sewage systems and extensive damage to roads and pavements. Thus, it is of great importance to find ways to control any damage processes in these systems.

1.2 Chemical and Physical Weathering

1.2.1 Chemical Weathering

A frequent and common form of chemical weathering is the sulfates attack on concrete, where cured concrete in the presence of moisture can be attack susceptible by sulfates. Those sulfates may be present in the water or from other sources. Attack occurs when the sulfates are able to react with the free lime released during hydration of the *Portland* cement and with calcium aluminates present in the cement. This reaction results in the formation of a range of sulfate compounds including gypsum and ettringite which is a hydrous calcium aluminum sulfate with formula: $Ca_6Al_2(SO_4)_3(OH)_{12} \cdot 26H_2O$. Because these compounds occupy a greater volume than the original concrete compounds, they cause expansion and the eventual failure of the concrete. Typical sulfate-resistant cements are considered to be low calcium aluminates (C_3A) such as *Portland* cements type V according to ASTM specification and blended Portland cements containing pozzolan or slag because pozzolanic materials, slag, and fly ash additions neutralizes portlandite and prevents the formation of ettringite, see Eqs. 1.1 and 1.2.

$$Ca(OH)_2 + Na_2SO_4 + 2H_2O \rightarrow CaSO_4 \cdot 2H_2O + 2NaOH \tag{1.1}$$

$$3CaO \cdot Al_2O_3 + 3(CaSO_4 \cdot 2H_2O) + 26H_2O$$
$$\rightarrow 3CaO \cdot Al_2O_2 \cdot 3CaSO_4 \cdot 32H_2O \tag{1.2}$$

1.2.2 Physical Weathering

Physical weathering is caused by the effects of changing temperature on rocks (concrete too), causing the rock to break apart. The process is sometimes assisted by water. There are two main types of physical weathering:

First – freeze-thaw occurs when water continually seeps into cracks, freezes and expands, eventually breaking the rock apart.

Second – exfoliation occurs as cracks develop parallel to the land surface a consequence of the reduction in pressure during uplift and erosion. Concrete structures are slowly and progressively affected by various natural weathering actions such as cyclic thermal changes, alternate wetting and drying. Physical weathering usually occurs in situ, without movement, when movement is involved weathering is augmented by erosion, which involves the movement of rocks and minerals by agents such as water, ice, snow, wind, waves and gravity and then being transported and deposited in other locations.

1.3 Microbiological Weathering

1.3.1 Microbiological Weathering

In addition to physical and chemical degradation processes, the presence of water contributes to undesirable changes in the material properties resulting from the activities of living organisms, i.e., biodeterioration. Since microorganisms are ubiquitous in almost every habitat and possess an amazingly diversified metabolic versatility, their presence on building materials is quite normal. Often, they can infer that can be detrimental (loss of alkalinity, erosion, spalling of the concrete skin, corrosion of rebars).

Concerning microbiological weathering, most of the concrete hydraulic structures and sewage collection systems damage suffers this kind of damage as a result of a sequence of processes involving biochemical transformations of sulfates and

Fig. 1.1 Sewer pipe debris
provides fresh sulfates to
sediments in a pipe being
corroded. *1* concrete pipe, *2*
residual water, *3* sulfate
reducing bacteria, *4* gypsum
and ettringite debris

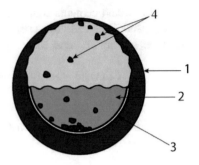

sulfur compounds in this material by the action of anaerobic and aerobic bacteria of the *Acidithiobacillus*, a genus of Proteobacteria. Like all Proteobacteria, *Acidithiobacillus* is Gram-negative. The members of this genus used to belong to *Thiobacillus*, before they were reclassified in the year 2000. With this genus the main deterioration tool is biogenic sulfuric acid. Several authors have described this problem [1–3]. The bacterial activity in the sewers creates a sulfur cycle, which could have led to the formation of sulfuric acid, when anaerobic conditions occur due to long residence time or the slow flow of the sewage and microorganisms start to populate the sewer [4] (See Fig. 1.1). Sulfate reducing bacteria, e.g., *Desulfovibrio*, reduce sulfur compounds to hydrogen sulfide (H_2S). All detrimental phenomena (malodors, concrete disintegration, health hazards) associated with hydrogen sulfide in gravity sewers depend on the rate of H_2S emission from the aqueous phase to the gas phase of the pipe. Due to heat, turbulence and pH decrease, H_2S escapes into the sewer atmosphere. Once in the atmosphere, it can react with oxygen to elementary sulfur, which is deposited on the sewer wall, where it becomes available as a substrate for oxidizing bacteria, *Thiobacilli* sp. [1]. Those bacteria convert the sulfur into sulfuric acid, which is used by the bacteria to dissolve the concrete and extract sulfates, see Eqs. 1.3 and 1.4.

$$Ca(OH)_2 + H_2SO_4 \rightarrow CaSO_4 \cdot 2H_2O \qquad (1.3)$$

$$3CaO \cdot Al_2O_3 + 3(CaSO_4 \cdot 2H_2O) + 26H_2O$$
$$\rightarrow 3CaO \cdot Al_2O_2 \cdot 3CaSO_4 \cdot 32H_2O \qquad (1.4)$$

Sulfates and sulfur compounds are considered essential in cement and concrete composition which can be broadly categorized as:

Added sulfates – gypsum, hemihydrates, anhydrite, several synthetic forms of sulfates (typically by-products like flue gas desulfurization materials); Clinker sulfates – including arcanite, aphthitalite, calcium langbeinite and thenardite.

Although normally reported as SO_3 (% by mass) for consistency, sulfur can be found in any combination of these forms. Elemental sulfur is almost never found in Portland cement, except in trace amounts, as it is normally produced in an oxidizing environment.

The added sulfates (gypsum) are blended with clinker during the finish grinding of the cement in amounts needed to control early setting properties, as well as shrinkage and strength development. The amount needed varies from cement plant to cement plant, depending on the chemistry and fineness of the cement, but is typically on the order of 5% by mass. The most common form of sulfate added to Portland cement is calcium sulfate ($CaSO_4$), some of which is intentionally dehydrated by the heat of grinding to form hemihydrates, which is more soluble and therefore available earlier to control early hydration reactions.

Sulfates form naturally during clinker production (sulfates typically are part of the raw materials as mined). The presence and effect of sulfates and sulfur compound in *Portland* cement are considered beneficial in such a degree that it has become an unbreakable paradigm that *Portland* cement has to contain them. Raw material sulfates tend to volatilize at the temperatures of cement kilns (up to about 1450 °C) and tend to condense on the outer surface of clinker nodules as alkali sulfates during the last stage of clinker production (rapid cooling). Furthermore, the amount depends on the chemistry of the raw materials and kiln operating conditions, making cement somewhat unique. These alkali sulfates are usually soluble enough to help control early hydration reactions. Some clinker sulfates are also incorporated into other cement production phases. Since cement is unique, chemical analyses are the best method of determining the SO_3 content of the final cements. Typically, the total SO_3 content is measured (or elemental sulfur is converted to SO_3 and measured) through methods in ASTM C114 (or AASHTO T 105). XRF analysis is probably the most common technique.

Consequently, almost every *Portland* cement is manufactured with some amount of sulfates. We strongly believe that the presence of sulfates in the concrete is the main reason of the biodeterioration process of sewers, see Fig. 1.1; the sulfates come from the concrete debris when being dissolved by the biogenic sulfuric acid. What's more, these debris rich in gypsum and ettringite lead to a significant increase of sulfates in the sewer sludge sediments closing a sulfur cycle in the sewer environment, pH lowers and concrete loses mass (see Fig. 1.2, after P. A. Wells [4]). Furthermore, sewage structures fill with bad odors and pollution gases resulting from sulfates reductions. Odors leak from slow flowing wastewater sewage. In the urban areas, these gases can also be the cause of the acid rain and affect building facades, and notably concrete external walls, can be affected by biological stains, which alter aesthetical quality of the construction, sometimes very quickly, and lead to significant cleaning costs.

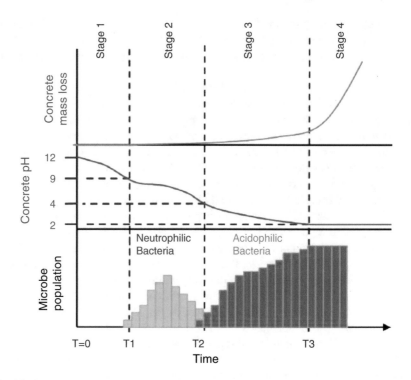

Fig. 1.2 Sewage conditions in relation with microbe population, concrete pH lowest and concrete mass lost (After P.A. Wells et al. [4])

1.3.2 Other Types of Microbiological Weathering

Weathering is in part carried out by other types of bacteria like cyanobacteria cyanophyte which are autotrophic and have characteristics of both algae and bacteria. They produce oxygen as a waste product of photosynthesis, such as algae, although they lack chloroplasts. They are single-celled organisms like bacteria, but most do not travel through flagella as bacteria do. Its blue-green color is due to the presence of chlorophyll and phycocyanin, therefore sometimes are known as blue-green algae. These microorganisms, algae and fungi responsible for these alterations have been quite well identified. Fungi are heterotrophic and lack chlorophyll, therefore, cannot make their own food and are dependent on organic matter available. Adhere to the substrate surface by means of hyphae (filamentous elements), usually hidden below the surface and appear as dark spots furry, gray, green, black or brown. Mold and yeast are included in this category. Fungi are immobile and spread depends on the release and transport of spores.

Lichen is a symbiosis of algae – green or blue green – and fungi; the fungal component covers most of the body. The nutrient components are synthesized by the algae, some of which are absorbed by the fungus, while the fungal component absorbs water through the body. Lichens can survive in a wide range of environmental

Fig. 1.3 If the conditions are right, lichen will consume and grow on the concrete as on any other soil

conditions because of their different appearance; in severe cases they may simply remain dormant instead of disappearing. Many lichens consist of fungal and photosynthetic symbionts that cannot live on their own. Others are facultative, meaning that they can, but do not have to live with the other organism. Lichens can burrow up to several millimeter deep into the surface of the stone. Adhere to the surface of the substrate by hyphae (filamentous elements) or roots with a preference for limestone. These symbionts are multicellular organisms, exceptionally resistant to adverse environmental conditions and able therefore to colonize diverse ecosystems. Protection against desiccation and solar radiation provided by the fungus and the photosynthetic capacity of the algae gives the symbiote unique features to this living being. Lichen substances make better use of water and elimination of harmful substances, see Fig. 1.3.

1.3.3   Sulfur and Limestone Content in Concrete Are the Main Reasons for Weathering Damage

Apparently the presence of sulfates, sulfur compound and limestone in cement and concrete is the main cause of the microbiologically induced weathering of concrete [5].

Most of the information in the literature does not provide a way to explain it, nor the information necessary to develop an accurate model that could be used to predict the lifecycle of the concrete nor be used to determine maintenance and repair schedules.

Every study has been directed to characterize and count the many types of microorganisms that populate the residual waters and inner space of a sewer infrastructure.

Table 1.1 Composition and properties of samples

	CLI Mix # 1	CLI Mix # 2	CPP Mix # 3	CPC Mix # 4	CPC Mix # 5	CPC Mix # 6
Cement (g)	740	740	667	740	2204	2775
Clinker (g)						
Microsilice (g)			74	70		
Talc (g)	70					
Silica sand (g)	2035	2035	2035	2035	2031	
Total (g)	2775	2775	2775	2775	2775	2775
Water in theory (ml)	382	382	384	380	1138	1450
w/c	0.52	0.52	0.58	0.52	0.52	0.52
Slump test (mm)	107.0	109.5	112.0	109.5	139.0	108.0
Water real (ml)	410.2	378.0	593.2	450.0	650.0	961.8
w/c real	0.55	0.51	0.89	0.59	0.29	0.35
f'c 3 days (MPa)	30.54	28.42	23.55	24.82	62.08	39.54
f'c 7 days (MPa)	35.89	31.28	26.97	34.14	72.08	77.86
f'c 28 days (MPa)	42.85	39.81	37.07	36.49	98.65	97.09

CLI clinker, *PPC* pozzolanic Portland cement, *CPC* composite Portland cement, class 40

Table 1.2 Cycles of dry-wet during hydration process and f'c (MPa) of samples

Samples	Mix # 1	Mix # 2	Mix # 3	Mix # 4	Mix # 5	Mix # 6
f'c (MPa) 28 days	42.85	39.81	37.07	36.49	98.65	97.09
f'c (MPa) 50 cycles	41.75	37.28	35.27	35.39	95.17	95.11
f'c (MPa) 100 cycles	40.46	35.35	33.16	33.47	93.33	93.28
f'c (MPa) 150 cycles	39.89	35.28	33.97	33.14	92.08	92.86
f'c (MPa) 200 cycles	40.15	35.81	33.07	34.49	92.65	92.09

1.4   Experimental

We prepare several sample of concrete, recorded in Table 1.1, to test its behavior during the process of hydration of *Portland* cement; significant amount of CaO is available as free lime, which, under cyclic wetting and drying is leached out in the form of $Ca(OH)_2$. This slow but progressive leaching consequently results in the development of secondary porosity and subsequently causes the weakening of the concrete structures.

This weathering process on concrete structures results from the action of rain and sun, but in the case of concrete hydraulic structures like canals, dams, etc. a similar effect is produced by the fluctuating flow of water.

Differently designed and cured specimens were subjected to 50, 100, 150 and 200 cycles of alternate wetting and drying in laboratory under simulated and controlled conditions. The extent of weathering suffered by these specimens has been assessed in terms of the loss of compressive strength of these specimens (See Table 1.2). A scrutiny of the results reveals that plain concrete specimens suffer loss in compressive strength and flexural strength due to cyclic wetting and drying action. It is interesting to note that the effect of induced weathering under different simulated conditions on laboratory specimens is influenced by the following factors: Richness of the mix, period of curing, water-cement ratio and presence of pozzolan.

Nonetheless, seems that sulfur, sulfates, limestone, calcite and calcareous components content in the concrete structures is the main reason of the damage, during the weathering process. In our experiment, the submerged surface of a sample is typically coated with a film comprised of a diverse microbial community. Anaerobic conditions can develop in such films, providing a suitable environment for the growth of sulfate-reducing bacteria. Sulfate reducers oxidize organic acids and alcohol, generated as end products of many types of anaerobic fermentation, with the concomitant reduction of sulfate to sulfide.

Several species capable of reducing and oxidizing sulfur compounds colonize exposed concrete. Anaerobic (desulfurization) microbial processes in the concrete sewer pipe lead to the formation of hydrogen sulfide, which is released to the inner pipe atmosphere. Part of the sulfide diffuses out of the film into the bulk liquid flow. At the pH of most wastewater, aqueous sulfide partitions between hydrosulfide (HS^-) and dissolved hydrogen sulfide (H_2S). Dissolved H_2S readily evolves into the sewer inner space, subsequently reaching the pipe crown see Fig. 1.1.

On the crown surface of the sewer pipe, hydrogen sulfide has reacted with oxygen, converted and deposited as elemental sulfur, throughout chemical oxidation under aerobic conditions; there sulfur and H_2S can be metabolized to sulfuric acid by sulfur-oxidizing bacteria, bacteria of the genus *Acidithiobacillus* which oxidizes them to sulfuric acid. The bacteria uses this acid to dissolve calcium hydroxide and calcium carbonate (when present) in the cement binder [6]; during the corrosion process gypsum and ettringite are deposited on the surface of the concrete and weaken the structural integrity of the concrete. This reduces and compromises the structural integrity of the concrete; the affected load-bearing capacity of the concrete can result in the eventual collapse of the concrete structure. While the concrete is slowly disintegrating, debris fall to the bottom of the pipe enriching the sediments with fresh sulfates, which are used (starting a new cycle) by sulfate-reducing bacteria at the bottom of the pipe, see Fig. 1.1.

In another environment when the amount of calcite in the concrete is high and the amount of intragranular and superficial water increases, providing the right conditions for other microorganisms to grow, the concrete loses cementitious components leaving the aggregate exposed, see Fig. 1.1.

During the sulfuric acid corrosion experiment the samples were hanging two inches above the water level of a sewer drain using a polypropylene cable resistant to biogenic sulfuric acid. We exposed the samples to this environment for 7 months; Table 1.3 shows the final mass loss of the samples.

Table 1.3 Mass lost experiment

		CLI	CLI	PPC	CPC	CPC	CPC
	(g)	Mix # 1	Mix # 2	Mix # 3	Mix # 4	Mix # 5	Mix # 6
Sample 1	Mass initial	265.81	268.54	258.03	276.24	284.46	248.80
	Mass final	265.11	266.15	240.19	265.43	272.81	231.55
	Lost	0.70	0.89	17.84	10.81	11.65	17.25
	%	0.26%	0.33%	3.04%	3.91%	4.10%	6.93%
Sample 2	Mass initial	265.97	267.32	260.40	248.80	283.91	
	Mass final	264.76	265.61	250.35	230.55	265.11	
	Lost	1.21	1.71	10.05	18.25	18.8	
	%	0.45%	0.63%	3.86%	7.34%	6.62%	
Average lost %		0.35%	0.48%	3.45%	5.62%	5.36%	6.93%

CLI clinker, *PPC* pozzolanic Portland cement, *CPC* composite Portland cement, class 40

1.5   Conclusions

We were not expecting lost of mass in the samples made without sulfur. Nonetheless, there was a small lost of mass in these samples. We believe the mass lost was because the consortium of *Acidithiobacillus* bacteria also contains *Acidithiobacillus ferrooxidans* that dissolves iron minerals which are a common constituent of *Portland* cement.

References

1. Parker CD (1945a) The corrosion of concrete 1. The isolation of a species of bacterium associated with the corrosion of concrete exposed to atmospheres containing hydrogen sulfide. Aust J Exp Biol Med Sci 23:81 pp
2. Sand W, Bock E (1984) Concrete corrosion in the Hamburg sewer system. Environ Technol Lett 5:517–528
3. Davis J, Nica D, Shields K, Roberts DJ (1998) Analysis of concrete from corroded sewer pipe. Int Biodeterior Biodegrad 42:75–84
4. Wells PA et al Factors involved in the long term corrosion of concrete sewers, Centre for Infrastructure Performance and Reliability, The University of Newcastle, Australia. Boon, A.G. Septicity in sewers: causes, consequences and containment. Wat Sci Tech 31(7):237–253
5. Rendon Diaz Miron LE (2010) Raw mix for the production of Portland cement clinker microbiological corrosion resistant. Patent Title # 282541 issued November the 24 of 2010, México City, México. http://www.pymetec.gob.mx/patentex.php?pn_num=MX0008444&pn_clasi= A&pn_fecha=2002-03-12
6. Rendon LE, Lara ME, Rendon M. The importance of *Portland* cements composition to mitigate sewage collection systems damage. http://dx.doi.org/10.1557/opl.2012.1547

Chapter 2
Influence of Sulfur Ions on Concrete Resistance to Microbiologically Induced Concrete Corrosion

Luis Emilio Rendon Diaz Miron and Maria Eugenia Lara Magaña

Abstract Concrete biodeterioration is defined as the damage that the products of microorganism metabolism, in particular sulfuric acid, do to hardened concrete. The combination of sulfur compounds and sulfur-dependent microorganisms is the origin of the process, because sulfates are found in certain groundwater, sewer and in sea water, additionally, some sulfur compounds are natural constituents of Portland cement. Along with this, the common presence and activity of microorganisms plays a very important function in the whole spectrum of degradation processes such as biodeterioration of metals and concrete. We report here the development of a possible biodeterioration resistant concrete. We assume that the elimination of sulfur compounds and acid reactive materials in the Portland cement and aggregates will prevent the formation and action of the biogenic acids that cause dissolution of calcium-containing minerals [for a narrative in Spanish see (Rendon, ¿Que es el biodeterioro del concreto? Revista Ciencia de la Academia Mexicana de Ciencias, Vol. 66 Num. 1. http://www.revistaciencia.amc.edu.mx/images/revista/66_1/PDF/Biodeterioro.pdf, 2014)]. This study was carried out on site inside a wastewater sewer drainage in Mexico City. Concrete samples whose main characteristic was being formulated without any sulfur or sulfates in its composition as well as reference concrete samples made with Ordinary Portland Cement (OPC) were used for the experiment. The weight changes and surface changes of both concrete samples were valuated after 7-month exposition to the biodeterioration process. The results obtained on site suggest that both the composition of concrete and duration of aggressive environment are very important. This possible biodeterioration-resistant concrete could give a viable solution to the long known problem of microbiologically induced concrete corrosion (MICC) a typical case of biodeterioration. Furthermore, we recommend the *Portland* type cement Mexican norm (ONNCCE,

L.E. Rendon Diaz Miron (✉)
Universidad Internacional (UNINTER),
304 San Jerónimo, Cuernavaca City, Morelos, Mexico, C.P. 62179
e-mail: luisrendon@uninter.edu.mx

M.E. Lara Magaña
Marudecori Consultants, 83 Cañada, Cuernavaca City, Morelos, Mexico, C. P. 62180

© Springer International Publishing AG 2017   11
L.E. Rendon Diaz Miron, D.A. Koleva (eds.), *Concrete Durability*,
DOI 10.1007/978-3-319-55463-1_2

Organismo Nacional de Normalización y Certificación de la Construcción y Edificación, S.C. (2004) Norma Mexicana (NMX – C – 414 – ONNCCE - 2004) Diario Oficial de la Federación 27 de julio de 2004, 2004) which does not take into consideration the concrete biodeterioration variable and its mechanism, to be reviewed in this aspect, or at least that a warning be issued as a key measure to mitigate biodeterioration in sewer concrete infrastructure.

Keywords Biodeterioration • Concrete • Corrosion Resistance

2.1 Introduction

Portland cement is obtained by a conventional process of sintering (solid state synthesis), from a mixture of limestone and clay to high temperature (1450 °C), where a partial melting of raw materials and formation of nodules of clinker are produced. The clinker is mixed with gypsum and grind together to make the cement. Adding sulfates and sulfur compounds are considered essential in cement and concrete composition to help with setting properties, lower the clinkering synthesis temperature, etc. Sulfur compound in cement and concrete can be broadly categorized as: Added sulfates—gypsum, hemihydrates, anhydrite, several synthetic forms of sulfates (typically by-products like flue gas desulfurization materials); Clinker sulfates—including arcanite, aphthitalite, calcium langbeinite, and thenardite. Although generally reported as SO_3 (% by mass) for consistency, sulfur can be found in any combination of these forms. Elemental sulfur is almost never found in *Portland* cement, except in trace amounts, as it is generally produced in an oxidizing environment. The added sulfates (gypsum) are blended with clinker during the finish grinding of the cement in amounts needed to control early setting properties, as well as shrinkage and strength development. The amount needed varies from cement plant to cement plant, depending on the chemistry and fineness of the cement, but is typically on the order of 5% by mass. The most common form of sulfate added to *Portland* cement is calcium sulfate ($CaSO_4 \cdot 2H_2O$), some of which is intentionally dehydrated by the heat of grinding to form hemihydrates, which is more soluble and therefore available earlier to control early hydration reactions, but sometimes may cause anomalous set as false set. Clinker sulfates form naturally during clinker production (sulfates typically are part of the raw materials as mined or fuel). These sulfates tend to volatilize at the temperatures of cement kilns (up to about 1450 °C) and tend to condense on the outer surface of clinker nodules as alkali sulfates during the last stage of clinker production (rapid cooling). Furthermore, the amount depends on the chemistry of the raw materials and kiln operating conditions, making cement somewhat unique. These alkali sulfates are usually soluble enough to help control early hydration reactions. Some clinker sulfate is also incorporated into other cement phases. Since cement is unique, chemical analyses are the best method of determining the SO_3 content of cements. Typically, the total SO_3

content is measured (or elemental sulfur is measured and converted to SO_3) through methods prescribed in ASTM C114 (or AASHTO T 105). XRF analysis is probably the most common technique. The presence and effect of sulfates and sulfur compound in cement and concrete are considered beneficial in such a degree that it has become an unbreakable paradigm, which *Portland* cement has to contain them. Consequently, every *Portland* cement is manufactured with some amount of them, therefore whatever the source of sulfates and sulfur compound in cement and concrete its presence is a fact.

2.2   Background

Most deterioration processes involve two stages. Initially, aggressive fluids (e.g., water, ionic solutions with dissolved salts, bacteria) need to penetrate or be transported through the capillary pore structure of the concrete to reaction sites (e.g., sulfates penetrating to reactive aluminate, bacteria penetrating to sulfur source gypsum) prior to the onset of the actual chemical or biochemical deterioration reactions. These fluids can be water, dissolved chemicals as sulfates, chlorides, alkalis, or acids (biogenic acids).

These aggressive fluids can be transported into or through the concrete pore structure by various single or combined mechanisms, some of which are:

Ionic diffusion (due to concentration gradient)
Vapor diffusion (due to humidity gradient)

2.2.1   *Sulfate Attack on* **Portland** *Cement Elements*

Sulfates will attack the calcium hydroxide, tricalcium aluminate (C_3A), and hydrated aluminate phases from the Portland cement. The calcium hydroxide can convert to gypsum:

$$Ca(OH)_2 + SO_4^= aq. \rightarrow CaSO_4 \cdot 2H_2O$$

However, gypsum formation is only a precursor to the attack.

The majority of the damage occurs through subsequent growth of ettringite crystals:

$$3CaSO_4 \cdot 2H_2O + C_3A + 25H_2O \rightarrow 3CaSO_4 \cdot C_3A \cdot 31H_2O$$

In part related to the large number of attached water molecules, the ettringite crystals are large and cause a volume increase, resulting in tensile forces and cracking of the concrete. Therefore, external sulfate attack results in cracking, softening, and exfoliation.

Usually, the $Ca(OH)_2$ content of the paste will be reduced through pozzolanic reaction and therefore will not be available to form gypsum or ettringite crystals as sulfate ions penetrate.

2.2.2 Microbiologically Induced Concrete Corrosion

There is evidence [14, 16] that sulfur compounds present in the concrete are the main reason for the dissolution of concrete by sulfate-reducing bacteria (SRB) through the microbiologically induced concrete corrosion (MICC) process. Corrosion of concrete has an enormous economic impact, especially when replacement or repair of corroded municipal sewer systems (sewer-pipe failure causes extensive damage to roads and pavements) or highway infrastructures is required.

Corrosion of concrete sewer pipes was detected as early as 1900 by Olmstead and Hamlin [7]. Hydrogen sulfide, an anaerobic decomposition product of sulfur-containing cement, was identified as one of the first corrosion perpetrator. Later, in a series of studies, Parker and coworkers established that the presence of *acidophilic thiobacilli* was correlated with concrete degradation [9–11]. Other important contributions in the microbial processes involved in concrete pipe degradation were done by Sand and Bock [20], Cho and Mori [1], Davis et al. [2], Nica et al. [6], Hernandez et al. [3], and Roberts et al. [17].

The biodeterioration of concrete sewers is the end result of a sequence of processes involving biochemical transformations of organic and inorganic sulfur compounds, by the action of anaerobic and aerobic bacteria. The submerged surface of a sewer pipe is typically coated with a film comprised of a diverse microbial community. Anaerobic conditions can develop in such films, providing a suitable environment for the growth of sulfate-reducing bacteria. Sulfate reducers oxidize organic acids and alcohol, generated as end products of many types of anaerobic fermentation, with the concomitant reduction of sulfate to sulfide [21]. Sulfide diffuses out of the film into the bulk liquid flow. At the pH of most wastewater, aqueous sulfide divides involving hydrosulfide (HS^-) and dissolved hydrogen sulfide (H_2S). Dissolved H_2S readily evolves into the sewer inner space, subsequently reaching the pipe crown. There the gas can be metabolized to sulfuric acid by sulfur-oxidizing bacteria. The sulfuric acid dissolves calcium hydroxide and calcium carbonate in the cement binder; then bacteria can reach inner cement sulfates a new source of sulfur, increasing corrosion and compromising the structural integrity of the pipe [19], see Fig. 2.1.

The microbial environment at the concrete surface is three-dimensional. Concrete is porous, and often it ends up coated by a porous layer of corrosion product (Fig. 2.3). Compounds and organisms are distributed throughout a volume that extends beneath the exposed surface. Dissolved sulfide at the surface diffuses into the concrete in both liquid and gas phases. Microbial oxidation of sulfide occurs throughout the depth of penetration. Sulfuric acid diffuses inward, dissolving and neutralizing alkaline substances as they diffuse outward from the underlying concrete, leaving an expanded residue of $CaSO_4$ on the surface, see Fig. 2.2.

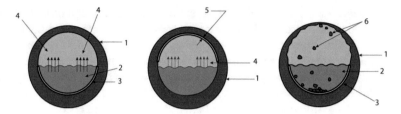

Fig. 2.1 Biodeterioration mechanism illustration: (*1*) concrete pipe, (*2*) wastewater stream, (*3*) anaerobic biofilm with sulfate-reducing bacteria, (*4*) internal atmosphere saturated with H_2S, (*5*) biofilm with sulfur-oxidizing bacteria (SOB), (*6*) concrete debris with gypsum and ettringite remains that fall to the bottom of the pipe and enrich the silt with sulfates [14]

Fig. 2.2 Expanded residue of $CaSO_4$ on the concrete surface

Fig. 2.3 Sulfur deposits at the visiting manholes of experimental site

New concrete has a low permeability, and only a small fraction of its pores are large enough to be penetrated by microorganisms. However, with time small interconnected cavities allow dissolved compounds to diffuse inward. Chemical gradients into the concrete are established quickly, porosity of the concrete increases as the dissolution of calcium hydroxide and calcium carbonate by acid enlarges the pores. With this puffy structure, the corrosion process is then further accelerated by the penetration of the concrete by microorganisms. Microbiologically induced concrete corrosion has been studied extensively, this concrete corrosion processes have been linked to the generation of sulfuric and nitric acids by sulfide oxidizing bacteria and nitrifying bacteria, nonetheless, the composition of concrete and presence of sulfates in it has not been implicated in the degradation of concrete. Furthermore, the objectives of this study were to evaluate the biodegradation of concrete whose composition is sulfates and sulfur compounds free in the binder cement and acid reactive compounds free in the aggregates.

2.3   Process Description

In the corrosion processes acid production is dominated by selected members of the genus *Thiobacillus thiooxidans* [4], which were recently reclassified as *Acidithiobacillus* by Kelly and Wood [5], and *Thiobacillus ferrooxidans* and the entire *Acidiphilium* genera [3, 13]. It has been known since 1945 [9] that *Acidithiobacilli* is the one that causes the rapid degradation of concrete sewage works and that corrosion occurred wherever *Acidithiobacillus* is present at an abundance of five x 10-exp. 6 cell/mg of total protein. Early work by Parker [12] as well as investigators at the Hamburg Municipal Drainage Department (Sand and Bock 1984) found *A. thiooxidans*, *A. neapolitanus*, *A. intermedius*, and *A. novellus* on sewer crowns. One approach to a better understanding of the microbial communities responsible for the production of sulfuric acid is through the analysis of their DNA. Sequence analysis of 16S rDNA and functional gene clone libraries (collections of genetic signatures) can be used to describe the phylogenetic affiliation (specific bacterial species) and function potential (what they can do) of microbial communities associated with MICC [3]. The development of comprehensive databases for bacterial communities in sewer systems within a wide range of corrosion conditions is crucial to identifying bacterial groups that could form the basis for bioassays used to develop innovative condition assessment tools for monitoring MICC.

Investigations of the causes of biogenic sulfuric acid corrosion of concrete particularly in the sewer pipes follow three criteria to diagnose MICC at the sewers: low pH measurements, the detection of neutrophilic sulfur bacteria in the pipes, and, associated with that, the observation of sulfur deposits at the visiting manholes, see Fig. 2.3. In a remarkable report, Rohwerder and Sand [18] proposed a biochemical model for elemental sulfur and sulfide oxidation in *Acidithiobacillus*. In this model, free sulfide is oxidized to elemental sulfur by a separate sulfide: Quinone oxidoreductase, and because this enzyme has only been isolated from neutrophilic sulfur bacteria, the detection of these types of bacteria ensures that the main corrosion mechanism, MICC, is underway. In the manufacture of concrete sewer pipes, microbiologically induced concrete corrosion phenomena are rarely taken into account. Durability considerations for the pipe material are as significant as its ability to perform intended structural and hydraulic functions. The capability of the pipe to continue to perform satisfactorily for an economically acceptable period is a fundamental engineering consideration. Unfortunately, predictions of durability cannot be made with the same degree of precision as can structural and hydraulic performance, and in too many instances, durability is not accorded adequate consideration. Furthermore, at the present time, no material is known that is completely inert to chemical or biochemical action and immune to physical deterioration. Concrete is no exception, but, under what might be considered *normal* exposure conditions, it has a very long life. Concrete made by the Romans from natural cement is in excellent condition after more than 2000 years of service.

2.4   Experimental

In this study it was investigated whether eliminating any sulfur compound from the concrete can prevent microbiologically induced concrete corrosion under realistic conditions of aggressive waters, e.g., residual or sewer waters. Another approach and a more elaborated discussion on this topic to prevent acid attack on concrete can be found in Rendon [15]. Furthermore, the study of more and different concrete compositions is under way. In this study the experimented concrete mixtures were composed of Ordinary Portland Cement (OPC), Pozzolanic Portland Cement (PPC), and Grinding Clinker Cement (GCC) as cement without sulfur, all mixed with silica sand, water, talc in sample 2 and fly ash in sample 6, see Table 2.1.

Specific factors that influence concrete pipe durability are concrete compressive strength, density, absorption, cement content (composition) and type, aggregate characteristics, total alkalinity, concrete cover over the reinforce, and admixtures.

To manufacture concrete when any kind of chemical deterioration is anticipated, it is convenient to go through Mexican norm (NMX-C-414-ONNCCE-2004) basic recommendations:

1. Utilize cement (type, SR) that is resistant to sulfate attack.
2. Utilize cement with mineral additions that have pozzolanic activity. This means that the hydrated products of the cement, the portlandite $Ca(OH)_2$ that would normally form, does not form because the mineral addition combines with the lime (CaO) to form a hydrated calcium silicate (C-S-H) product. This gives the characteristics of resistance, durability, and impermeability to the concrete. It avoids the formation of portlandite, which is leachable and has less resistance and durability.
3. Reduce the water/cement ratio to less than 0.45% (a fraction of 1%) maximum. This reduces the permeability of the concrete and limits the access of aggressive

Table 2.1 Concrete samples composition and data

Concrete						
	OPC 40	GCC	GCC	PPC 30	OPC 40	OPC 40
Sample composition	Sample 1	Sample 2	Sample 3	Sample 4	Sample 5	Sample 6
Cement	740 g			740 g	2775 g	600 g
Clinker		2204 g	740 g			
Fly ash						140 g
Talc		70 g				
Silica sand	2035 g	501 g	2035 g	2035 g		2035 g
Total sulfur[a] SO_3%	4.2%	1.0%	0.8%	4.3%	3.5%	3.7%
Water in theory	382 g	1138 g	382 g	382 g	1450 g	384 g
w/c	0.52	0.52	0.52	0.52	0.52	0.58
Slump	107	139.5	139	109.5	108	112
Water real	410.2 g	650 g	405 g	378 g	961.8 g	593.2 g
w/c real	0.55	0.29	0.55	0.51	0.35	0.89
f'c 3 days MPa	30.54	54.82	42.08	25.42	39.54	20.55
f'c 7 days MPa	35.89	64.14	45.08	31.28	77.86	26.97
f'c 28 days MPa	42.85	76.49	58.65	39.81	97.09	37.97
Initial sample weight	318.5 g	342.5 g	318.0 g	315.3 g	373.7 g	336.8 g
Final sample weight	222.9 g	332.2 g	311.6 g	223.9 g	291.4 g	252.6 g
Mass lost %	30%	3%	2%	29%	22%	25%

[a]Each composition was analyzed in a Sulfur Analyzer – The LECO Corp. Model CS-225 induction furnace

fluids into the interior of the mass. The implementation of these three recommendations requires an understanding of the microstructure of the cement paste that unites stony aggregates, its density (diminished porosity), and the process of concrete manufacture including the curing process.

In addition to these recommendations our concrete samples were made with sulfur-free *Portland* cement and limestone-free aggregates (sand and coarse aggregates) or any other acid reactive mineral. So we:

(a) Eliminate any source of sulfur from the cement and concrete.
(b) Eliminate any acid reactive mineral from the aggregates.

See Table 2.1.

2.4.1 Concrete Composition

Concrete is a manufactured material, so the answer to its degradation problem shall focus on why concrete is being corroded? The answer to this question is concrete corrodes because it contains a significant amount of sulfur compounds. Nonetheless,

Fig. 2.4 Samples
reviewed after a couple of
months in experimental
site

we know that correcting the composition of *Portland* cement and concrete is difficult, and complicated for two reasons: first, cement manufacturers are convinced
that sulfur is beneficial for cement, and second, to remove all sulfur compounds
from raw material can be difficult and costly. To mitigate biodeterioration eliminate
or minimize sulfur content or look for a substitute. Nonetheless, MICC is a fact
Fig. 2.4 shows the aspect of some of the different composition samples after only
2-month exposition to the biodeterioration process.

2.5   Results

Table 2.1 shows that in the concrete samples without sulfates mass lost was very
low, despite the fact that all samples were settled down under the same conditions.
The sulfuric acid responsible for the corrosion of concrete sewer systems is generated by a complex microbial ecosystem. Several species capable of oxidizing sulfur
compounds colonize exposed concrete. When sulfates are present anaerobic microbial processes in the sewage lead to the formation of hydrogen sulfide, which is
released to the atmosphere through turbulence. On the crown surface of the sewer
pipe, hydrogen sulfide is chemically oxidized to sulfur under aerobic conditions;
bacteria of the genus *Acidithiobacillus* oxidize sulfur to sulfuric acid (Fig. 2.1) dissolving the concrete, and losing mass while searching for more sulfates. Furthermore,
if the cement of the concrete contains large amounts of calcite as filler, it will be
easily dissolved by the sulfuric acid; more mass will be lost making the concrete
porous, with porous layers of corrosion product coating it, Fig. 2.2.

2.6   Discussion

The expected outcomes of this research are lowered costs of operation, maintenance
and replacement of aging water infrastructures, and reduced life-cycle costs of
wastewater conveyance systems. Preliminary results of this experimentation using

concrete without sulfates, sulfur compounds, and limestone show that mass lost by MICC effect is lowered. In Mexico, the *Portland* type cement recommended for use with aggressive waters, e.g., residual or sewer waters are usually Pozzolanic Portland Cement (PPC) and Granulated Slag Portland Cement (GSPC) which contains calcium sulfides, besides the added calcium sulfate. And lately, there has been an emphasis to use "Blended Portland Cement" (BPC) at the same level as PPC and GSPC, see the Mexican Norm NMX-C-414 [8]. This is a poor practice because BPC contains large amounts of calcite as filler that is easily dissolved by the biogenic acids allowing a greater concrete mass loss. In conclusion regarding the composition of concrete for sewer systems, the Mexican Norm NMX-C-414 [8] does not take into consideration the concrete biodeterioration variable and its mechanism, and its recommendation that blended cement be used in situations of aggressive chemical attack neglects the hazard that results from the incorporation of a significant amount of calcite. We recommend that the norm be reviewed, or at least that a warning be issued.

2.7   Conclusion

The importance of ensuring durability of concrete has been a growing concern of engineers, and there is now considerable understanding of the mechanisms, which cause its deterioration throughout non-living environment – chlorides in seawater, carbon dioxide in the atmosphere, cyclic freezing and thawing – and means of limiting such damage through the use of appropriate materials. However, many of the deterioration mechanisms, which affect concrete, are the result of interaction with living organisms which cause damage – through both chemical and physical processes – as was previously discussed and which under the right conditions, can be severe. We must conclude that both non-living environment mechanisms as well as living organism mechanisms must be taken into consideration to expand concrete durability.

References

1. Cho K, Mori T (1995) A newly isolated fungus participates in the corrosion of concrete sewer pipes. Water Sci Technol. 31:263–271
2. Davis J, Nica D, Shields K, Roberts DJ (1998) Analysis of concrete from corroded sewer pipe. Int Biodeterior Biodegrad 42:75–84
3. Hernandez MT, Marchand DJ, Roberts DJ, Peccia JL (2002) In situ assessment of active Thiobacillus species in corroding concrete sewers using fluorescent RNA probes. Int Biodeterior Biodegrad 49:271–276
4. Karavaiko GI, Pivovarova TA (1973) Oxidation of elementary sulfur by Thiobacillus thiooxidans. Microbiol Leningrad USSR 42(3):389–395

5. Kelly DP, Wood AP (2000) Reclassification of some species of Thiobacillus to the newly designated genera Acidithiobacillus gen. nov., Halothiobacillus gen. nov. and Thermithiobacillus gen. nov. Int J Syst Evol Microbiol 50:511–516
6. Nica D, Dickey J, Davis J, Zuo G, Roberts DJ (2000) Isolation and characterization of sulfur oxidizing organisms from corroded concrete in Houston sewers. Int Biodeterior Biodegrad 46:61–68.
7. Olmstead WM, Hamlin H (1900) Converting portions of the Los Angeles outfall sewer into a septic tank. Engineering news 44:317–318.
8. ONNCCE (2004) Organismo Nacional de Normalización y Certificación de la Construcción y Edificación, S.C. (2004) Norma Mexicana (NMX – C – 414 – ONNCCE – 2004) Diario Oficial de la Federación 27 de julio de 2004
9. Parker CD (1945) The corrosion of concrete 1. The isolation of a species of bacterium associated with the corrosion of concrete exposed to atmospheres containing hydrogen sulfide. Aust J Exp Biol Med Sci 23:81
10. Parker CD (1945) The corrosion of the concrete II. The function of Thiobacillus concretivorus nov spec in the corrosion of concrete exposed to atmospheres containing hydrogen sulfide. Aust J Exp Biol Med Sci 23:91–98
11. Parker CD (1947) Species of sulfur bacteria associated with the corrosion of concrete. Nature 159(4039):439–440
12. Parker CD (1951) Mechanisms of corrosion of concrete sewers by hydrogen sulfide. Sewage Ind Wastes 23(12):1477–1485
13. Peccia J, Marchand EA, Silverstein J, Hernandez MT (2000) Development and application of small-subunit rRNA probes for the assessment of selected Thiobacillus species and members of the genus Acidophilium. Appl Env Microbiol 66(7):3065–3072
14. Rendon LE, Lara ME, Rendon M (2012) The importance of Portland cement composition to mitigate sewage collection systems damage, Cambridge Journal Online. doi: http://dx.doi.org/10.1557/opl.2012.1547
15. Rendon LE (2013) Crude mixture for producing clinker, the subsequent production of Portland-type cement, and concrete resistant to a direct chemical acid attack. Publication WO/2013/191524, Dec 2013, México. https://patentscope.wipo.int/search/es/detail.jsf?docId=WO2013191524
16. Rendon LE (2014) ¿Que es el biodeterioro del concreto? Revista Ciencia de la Academia Mexicana de Ciencias. 66(1). http://www.revistaciencia.amc.edu.mx/images/revista/66_1/PDF/Biodeterioro.pdf
17. Roberts DJ, Nica D, Zuo G, Davis J (2002) Quantifying microbially induced deterioration of concrete: initial studies. Int Biodeterior Biodegrad 49(4):227–234
18. Rohwerder T, Sand W (2003) The sulfane sulfur of persulfides is the actual substrate of the sulfur-oxidizing enzymes from Acidithiobacillus and Acidiphilium spp. Microbiology 149:1699–1709
19. Sand W (1987) Importance of hydrogen sulfide, thiosulfate, and methylmercaptan for growth of Thiobacilli during simulation of concrete corrosion. Appl Environ Microbiol. 53:1645–1648
20. Sand W, Bock E (1984) Concrete corrosion in the Hamburg sewer system. Environ Tech Lett. 5:517–528
21. Smith DW (1992) Ecological actions of sulfate-reducing bacteria. In: Odom JM, Singleton R (eds) Sulfate-reducing bacteria: contemporary perspectives. Springer, Berlin, pp 161–188

Chapter 3
The Onset of Chloride-Induced Corrosion in Reinforced Cement-Based Materials as Verified by Embeddable Chloride Sensors

F. Pargar, Dessi A. Koleva, H. Kolev, and Klaas van Breugel

Abstract The need for an accurate determination of the chloride threshold value for corrosion initiation in reinforced concrete has long been recognized. Numerous investigations and reports on this subject are available. However, the obtained chloride threshold values have always been, and still are, debatable. The main concern is linked to the methods for corrosion detection and chloride content determination in view of the critical chloride content itself. In order to measure the chloride content, relevant to the corrosion initiation on steel, destructive methods are used. These traditional methods are inaccurate, expensive, time consuming and noncontinuous. Therefore, the application of a cost-effective Ag/AgCl ion selective electrode (chloride sensor) to measure the chloride content directly and continuously is desirable. The advantage would be an in situ measurement, in depth of the concrete bulk, as well as at the steel/concrete interface.

The aim of this work was to evaluate the importance of the sensor's properties for a reliable chloride content measurement. The main point of interest with this regard was the contribution of the AgCl layer and Ag/AgCl interface within the process of chloride content determination in cementitious materials. The electrochemical behavior of sensors and steel, both embedded in cement paste in a close proximity, hence in identical environment, were recorded and outcomes correlated towards clarifying the objectives of this work. The main point of interest was to simultaneously detect and correlate the time to corrosion initiation and the critical chloride content.

The electrochemical response of steel was monitored to determine the onset of corrosion activity, whereas the sensors' electrochemical response accounted for the

F. Pargar (✉) • K. van Breugel
Delft University of Technology, Faculty of Civil Engineering and Geosciences,
Department Materials and Environment, Stevinweg 1, 2628, CN, Delft, The Netherlands
e-mail: f.pargar@tudelft.nl

D.A. Koleva
Delft University of Technology, Faculty of Civil Engineering and Geosciences,
Section Materials and Environment, Stevinweg 1, 2628, CN, Delft, The Netherlands

H. Kolev
Institute of Catalysis, Bulgarian Academy of Sciences,
Acad. G. Bonchev str., bl. 11, 1113 Sofia, Bulgaria

© Springer International Publishing AG 2017
L.E. Rendon Diaz Miron, D.A. Koleva (eds.), *Concrete Durability*,
DOI 10.1007/978-3-319-55463-1_3

chloride content. For evaluating the electrochemical state of both sensors and steel, electrochemical impedance spectroscopy (EIS) and open circuit potential (OCP) measurements were employed. The results confirm that determination of the time to corrosion initiation is not always possible and straightforward through the application of OCP tests only. In contrast, EIS is a nondestructive and reliable method for determination of corrosion activity over time. The obtained results for corrosion current densities for the embedded steel, determined by EIS, were in a good agreement with the sensors' half-cell potential readings. In other words, the sensors are able to accurately determine the chloride ions activity at the steel/cement paste interface, which in turn brings about detectable by EIS changes in the active/passive state of steel.

The electrochemical response was supported by studies on the morphology and surface chemistry of the sensors, derived from electron microscopy (ESEM) and X-ray photoelectron spectroscopy (XPS). It can be concluded that the accuracy of the sensors, within detection of the time to corrosion initiation and critical chloride content, is determined by the sensors' properties in terms of thickness and morphology of the AgCl layer, being an integral part of the Ag/AgCl sensors.

Keywords Ag/AgCl electrode • Chloride sensor • Steel corrosion • Cement paste • Microstructural and Electrochemical tests

3.1 Introduction

The steel reinforcement in concrete is in passive state due to a thin iron oxide layer (passive film) formed on the steel surface in the conditions of this high pH medium (12.5 < pH <13.5) [1]. Exposure to marine environment and subsequent chloride ions penetration into the hardened concrete increases the chloride ion concentration above the threshold value for corrosion initiation on the reinforcing steel. This process locally destroys the passive film, originally present on the steel surface, and induces corrosion [2]. It is generally accepted that many factors affect the onset of localized corrosion. Among these, the quality of the steel-concrete interface and the pH of the pore solution are considered as the most influential parameters. Additionally, the level of chloride concentration at which depassivation occurs is of interest in view of predictive investigations on corrosion initiation and propagation. Judgment of the relevant chloride concentration, however, is largely affected by the chloride determination methods themselves [3, 4]. In other words, an ideal corrosion detection method and/or chloride content measurement technique should not intervene in a way to produce significant alteration at the interface steel/cementitious material and should not affect the steel surface properties.

A commonly used nondestructive electrochemical technique, to determine the time to corrosion initiation of embedded steel, is the open circuit potential (OCP) reading (or half-cell potential) [5]. With this approach, kinetic data on the rate of corrosion cannot be provided. Scattered data of OCP measurements from steel in concrete are frequently encountered due to variables that determine and influence the corrosion process. OCP measurements are affected by a number of factors, including the bulk cement matrix, e.g., additional resistance determined by bulk matrix porosity and permeability, the presence of highly resistive product layers, often limiting oxygen diffusion [3]. Thus, it is generally accepted that OCP records must be complemented by other methods. Although reliable relationships between potential and corrosion rate can be found in the laboratory for well-established conditions, these can in no way be generalized, since wide variations in the corrosion rate are possible in an otherwise narrow range of recorded potentials [6, 7].

For lab conditions, nondestructive sophisticated techniques, like Electrochemical Impedance Spectroscopy (EIS), can be employed and results coupled to the OCP tests. The combination of complementary techniques is then expected to result in a thorough evaluation of a reinforced concrete system. For example, EIS can simultaneously provide information on both steel electrochemical behavior and electrical properties of the cement-based bulk matrix, including properties at the steel/cement paste interface [8, 9]. EIS is nondestructive in the sense that an AC electrical signal, of sinusoidal wave with a small amplitude, is applied within measurement. This does not induce the accumulation of DC polarization, a phenomenon which would otherwise modify the steel surface. Therefore, EIS is suitable for identifying the onset of steel corrosion [6, 10, 11]. Additionally, judging from the recorded EIS response, the method is useful to indicate the presence of chloride ions in the vicinity of the steel surface, even prior to further corrosion propagation [12]. However, a quantitative assessment in view of deriving a chloride threshold value for localized corrosion initiation through EIS is not possible. This is especially the case if results from EIS tests are not supported or validated by quantitative assessment of the chloride ion concentration.

The combination of electrochemical techniques with a reliable chloride determination method would result in a significantly better detection and prediction of chloride-induced steel corrosion. Local chloride measurement at the depth of the embedded steel is preferable, rather than chloride determination at the concrete cover depth and/or the bulk concrete, which is the approach of traditional chloride determination methods. In other words, accurate and reliable chloride sensors would be the best solution in view of the above considerations. Although the principles of this approach are well recognized, sensors' application in the engineering practice is still limited.

Due to their small dimensions Ag/AgCl electrodes (chloride sensors) can be used for localized corrosion, hence more precise measurements, without changing the surrounding environment. The chloride ion concentration is "measured" at the depth of the embedded steel, rather than by averaging values, derived over the comparatively large concrete volume under investigation [13]. Ag/AgCl electrodes, as chloride sensors, have been studied for continuous nondestructive monitoring of the free chloride ion content in cement-based materials since 1993 [14]. They are predominantly sensitive to chloride ions and according to the Nernst equation exhibit a certain electrochemical potential that depends on the chloride ion activity (concentration respectively) in the environment. Although this principle is well known, the calculation of chloride concentration from the Nernst equation is not straightforward [12, 15–17]. A measurement error of a few millivolts adversely affects the accuracy of the method. Sources of errors can be linked to the specific microstructure and morphology of the AgCl layer and the Ag/AgCl interface, the sensor/cement paste interface characteristics and environmental conditions (temperature, alkalinity, etc.). These are in addition to the geometrical position of the sensors with respect to a reference electrode, which was discussed comprehensively and reported by Angst et al. [6, 15, 18].

Next to the above, well known is that chloride sensors exhibit certain limitation in high pH medium, as concrete. In these conditions, the sensor potential changes due to the gradual transformation of AgCl into Ag_2O. At 25 °C, the solubility product of AgCl and AgOH, and Ksp(AgCl) and Ksp(AgOH) is 1.8×10^{-10} and 2×10^{-8}, respectively [19]. Thus, it can be expected that when the hydroxide ion concentration is 100 times larger than the chloride concentration, the AgCl layer may transform into Ag_2O thermodynamically. Considering the pH value of 13.5 for concrete pore solution, the AgOH formation may occur at the chloride concentration below 3 mM. However, AgOH is unstable and tends to convert to Ag_2O. The sensor consisting of AgCl layer acts as a chloride sensor, but the sensor with Ag_2O layer is sensitive to hydroxide ions and thus the electrode acts as a pH sensor rather than a chloride sensor.

Additionally, in alkaline environment with low chloride concentration different electrode potentials for similar chloride sensors were reported. Among various phenomena, responsible for the reduced sensitivity and/or accuracy of the sensors, this lack of agreement was attributed to the methods of sensor preparation [13, 20]. However, no evidence to support this assertion was provided so far. Therefore, a more illustrative and in depth investigation is needed to shed light on the matter.

Within our previous studies it was shown that the properties of the AgCl layer affect the sensors' sensitivity, especially at low chloride concentration (tests were performed in simulated pore solution [21]). The AgCl layer, formed (or deposited) on the Ag surface, can vary in thickness, which depends on the preparation regime. As previously recorded, the thicker and more complex the AgCl layer is (e.g., more than one interface was observed within AgCl layers of above 20 μm thickness), the more time is needed for the sensor to "respond" to a low chloride concentration, especially in a high pH environment as concrete [21]. This is as expected, since the stable potential of the sensor is the consequence of electrochemical stability and equilibrium at the Ag/AgCl interface which is dependent on the morphology, composition and microstructure of AgCl layer [22].

This work will discuss the influence of the AgCl layer properties in view of critical chloride concentration detection (i.e., sensor response) at the time of steel corrosion initiation (i.e., steel response). For this purpose, sensors produced at different anodization regimes were embedded together with steel rods in cement paste cylinders (Fig. 3.1 further below). The cylinders were immersed in simulated pore solution (SPS), containing 870 mM chloride concentration. Monitoring the response of the steel rods aimed to evaluate the corrosion initiation and propagation stages, whereas monitoring the sensors' response aimed to "couple" these events with the relevant chloride content at the same time interval.   OCP values for both sensors and steel rods were recorded over 180 days. EIS was employed for qualification of

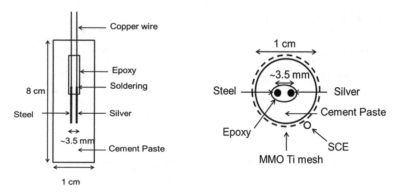

Fig. 3.1 Schematic representation of (**a**) the experimental setup and (**b**) electrochemical measurement configuration

the corrosion state of the steel rods. The paper correlates the electrochemical behavior of sensors and steel, recorded simultaneously in the same environment, which finally reflects the importance of the AgCl properties in view of the sensors' response, accuracy and reliability.

3.2   Experimental

3.2.1   Materials and Specimen Preparation

Silver wires (99.95% purity), 1 mm in diameter and 2.5 cm in height, were anodized at three different anodization regimes (Table 3.1) according to the following procedure: (1) the Ag wires were cleaned for 2 h in concentrated NH_4OH and immersed in distilled water overnight; (2) an exposed length of 1.5 cm was anodized in 0.1 M HCl for 1 h at different current densities (Table 3.1); (3) the anodized silver wire was soldered to a copper wire and the 0.5 cm of the soldered zone, together with the nonanodized part, protected with an epoxy resin; the final exposed length of the sensors was 1 cm. The produced sensors were studied prior to casting in the cement paste cylinders. The main points of interest were surface morphology and composition in view of the effect of the different anodization regimes (Table 3.1).

Steel wires and cement paste cylinders: steel wires (1 mm diameter), drawn from low carbon steel, were acetone-cleaned and epoxy-insulated except an exposed length of 1 cm. The sensor and steel rods were "coupled" and embedded in 1 cm diameter cement paste cylinders in which only 1 cm length of the steel and sensor were exposed to the environment, whereas the remaining parts were insolated by epoxy (Fig. 3.1). The active surface of the steel rods and the sensors was 0.39 cm^2 and 0.32 cm^2, respectively. The cement paste cylinders were cast using Ordinary Portland Cement OPC CEM I 42.5 N (producer ENCI, NL) at water-to-cement ratio of 0.4. After curing in a sealed condition for 30 days, the specimens were immersed

Table 3.1 Anodization regimes for sensor preparation

Regime	Current density (mA/cm^2)	Duration of the anodization (h)	Measured thickness of AgCl layer (μm)
A	0.5	1	6–10
B	1	1	~15
C	2	1	~20
D	4	1	~40

in a simulated pore solution (SPS) with the following composition: 0.05 M NaOH +0.63 M KOH + Sat. Ca(OH)$_2$. The pH of the SPS medium was maintained at 13.6.

The desired level of chloride concentration in the solution was adjusted to 870 mM by adding NaCl (as a solid). The container was covered to prevent evaporation. In order to achieve a relatively constant chloride concentration, the volume ratio of solution to cement paste was maintained at 40 throughout the full duration of the test.

3.2.2 Methods

The sensors' surface morphology was analyzed using Environmental Scanning Electron Microscopy (ESEM), Philips-XL30-ESEM, equipped with an energy dispersive spectrometer (EDS) at accelerating voltage of 20 kV and in high vacuum mode. The composition of the AgCl layers, obtained at different anodization regimes was evaluated through X-ray photoelectron spectroscopy (XPS). The measurements were carried out using an ESCALAB MkII (VG Scientific) electron spectrometer at a base pressure in the analysis chamber of 5×10–10 mbar using twin anode MgKα/ AlKα X-ray source with excitation energies of 1253.6 and 1486.6 eV, respectively. The XPS spectra were recorded at the total instrumental resolution (as it was measured with the FWHM of Ag3d5/2 photoelectron line) of 1.06 and 1.18 eV for MgKα and AlKα excitation sources. The processing of the measured spectra includes a subtraction of X-ray satellites and Shirley-type background [23]. The peak positions and areas were evaluated by a symmetrical Gaussian-Lorentzian curve fitting. The relative concentrations of the different chemical species were determined based on normalization of the peak areas to their photoionization cross sections, calculated by Scofield [24].

Open circuit potential (OCP) for both sensors and steel rods was monitored over time, versus a saturated calomel electrode (SCE). Electrochemical impedance spectroscopy (EIS) was performed for the steel rods at certain time intervals, using a three-electrode cell arrangement, where MMO Ti cylinders, positioned around the cement paste cylinder, served as counter electrode, the steel rods served as working electrode and a SCE electrode, immersed in the solution, served as a reference electrode (Fig. 3.1). EIS was employed using 10 mV AC perturbation (rms) in the frequency range of 50 kHz to 10 mHz. The equipment used was a Metrohm Autolab

PGSTAT 302 N, combined with a FRA2 module and GPES/FRA and NOVA software packages.

3.3 Results and Discussion

3.3.1 Morphology and Composition of the AgCl Layers

The lack of evidence-based practice for recorded various OCP values of Ag/AgCl electrodes (as chloride sensors), in alkaline environment of otherwise the same chloride content, is often a concern and subject to discussion [13]. For a better understanding of the hypothesized phenomena, responsible for these contradictory results, the morphology and microstructure of the as formed during anodization AgCl layers were studied via ESEM. These results, together with XPS qualification, support the recorded electrochemical response of the sensors in view of chloride content determination.

Considering the material properties of Ag and Ag-based compounds, sample preparation for studying the cross section of a Ag/AgCl system is challenging in many ways, from sample preparation to storage and handling. The following procedure was followed for ESEM observations: (i) a portion of the silver wire was narrowed prior to anodization; (ii) the narrowed portion was stretched from the two sides of the notch; (iii) this allowed microscopic investigation of a "cross section," i.e., the parallel growth of the AgCl layer on the Ag substrate. The obtained cross section was not smooth as a clean-cut surface, but rather similar to a fracture surface. This allowed the morphology and the inner distribution of the silver chloride particles to be well observed. It is important to note that, the attempts of other researchers to capture and describe the inner morphology of AgCl layers and the AgCl particles' distribution were so far not satisfactory and in many cases were speculative, rather than conclusive [25–27]. Therefore, this study is a contribution to the state of the art, by providing a clear evidence on the variation of both morphology (including thickness) and composition of AgCl layers, determined by the different anodization regimes. By raising current density from 0.5 mA/cm² (regime A) to 4 mA/cm² (regime D), the AgCl layer's thickness increased from 6 to 40 μm (Table 3.1). Results from an analytical approach to determine the AgCl layer thickness (5–37 μm based on applying Faraday's law [28, 29]) are well in line with the actually observed layers. The morphological features of these layers are as presented in Fig. 3.2. As can be observed, the thickness of the AgCl layer increased with increasing current density during anodization, as expected. The varying AgCl layer thickness was accompanied by a different morphology and packing of the layer, which again, depends on the applied current density. In regimes A and B, "packed-piled" AgCl particles were generated on the Ag substrate (Fig. 3.2a, b). The cross section of a AgCl layer in regime A showed densely packed particles, approx. 1–2 μm wide and ~6–10 μm in height (Fig. 3.2a). Increasing the current

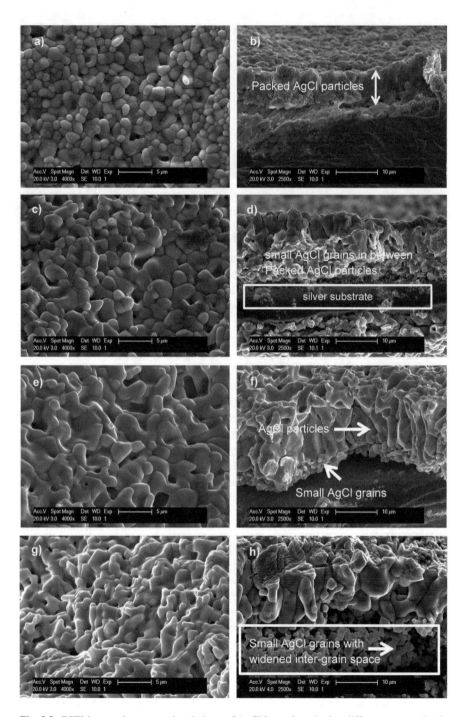

Fig. 3.2 ESEM top and cross-sectional views of AgCl layer deposited at different current densities; (**a**) top view and (**b**) cross section of sensor at regime A; (**c**) top view and (**d**) cross section of sensor anodized at regime B; (**e**) top view and (**f**) cross section of sensor anodized at regime C; (**g**) top view and (**h**) cross section of sensor anodized at regime D

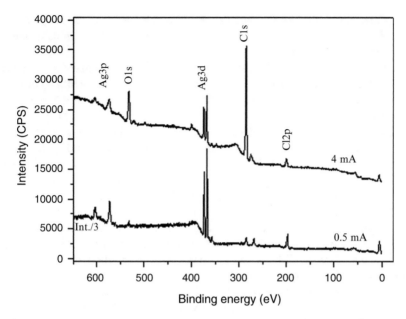

Fig. 3.3 Survey XPS spectra for sensors A and D, obtained at 0.5 and 4 mA/cm² anodization regimes – overlay for A and D sensors

density to 1 mA/cm² (regime B) resulted not only in a thicker silver chloride layer (~15 μm) but also induced the occurrence of smaller AgCl particles close to the silver substrate and in between the "packed-piled" AgCl particles (Fig. 3.2c, d). In contrast, in regimes C and D these features became a mosaic of complex patterns (Fig. 3.2e, g), in which more porous-discontinuous AgCl grains were detected close to the silver substrate (see arrows in Fig. 3.2f, h). In regimes C and D, the "piled" AgCl particles cannot be considered individually and instead of one AgCl layer a multilayered structure was observed (Fig. 3.2e, h). Small particles, separated from the "twisted" AgCl top layer with a borderline in between, were observed in the proximity of the Ag substrate (Fig. 3.2f, h – marked regions). At higher current density regimes, the AgCl particles in the vicinity of the Ag substrate were smaller, whereas their inter-grain channels were widened.

XPS analysis of the sensors provides information about the surface composition, i.e., a confirmation for AgCl layers as such, but also gives information for possible impurities. Typical XPS spectrums as survey-scans are depicted in Fig. 3.3 as an overlay for specimens A (0.5 mA regime) and D (4 mA regime). The survey-scans show dominant peaks for Ag3p, Ag3d and Cl2p, along with peaks for oxygen (O1s) and carbon (C1s) impurities.

Except overall information for chemical composition, the survey-scans in Fig. 3.3 can indicate quantitative variations. It can be well seen that for the different specimens, e.g., for sensors A and D of the lowest and highest thickness of the AgCl layer, the peaks of relevant elements are of different intensity. While the presence of Ag and Cl was as expected, the presence of oxygen and carbon can be denoted to contamination or impurities.

Although care was taken for minimizing the effect of external environment (O_2, CO_2, humidity), the sensors were in contact with atmosphere while handling in-between tests and transfer to vacuum chambers. What is interesting to note is that surface adsorption of substances from the environment was obviously larger for the thicker AgCl layers, as evident by the significantly higher peaks for C1s and O1s in regime D, if compared to regime A, Fig. 3.3. In other words, the presence of oxygen and humidity in the environment would be expected to affect more significantly sensors with a thicker AgCl layer. Except larger thickness and, therefore, a larger amount of adsorbed species, the higher level of impurities in thicker AgCl layers is most likely additionally affected by the multilayered morphology in these cases (Fig. 3.2), resulting in a larger "active" surface area for adsorption reactions to take place. Additionally, it is well possible that chemical recombination in depth of the AgCl layers was relevant, but qualification and quantification in depth of the layers cannot be judged from XPS as performed in this experiment. The results, however, clearly support the hypothesis of chemical transformations of the thicker AgCl layers. These are also in line with the microscopic observations of a multilayer AgCl structure and the effect of surface morphology on chloride content determination (sensors' response), which are to be discussed further below in Sect. 3.3.3.2.

To this end, a synergetic effect of the increasing thickness, roughness and multilayer structure of the AgCl for the case of B, C and D (as observed, Fig. 3.2) would account for a more pronounced chemical recombination and transformation of the AgCl layer (as recorded, Fig. 3.3) if compared to the case of A sensors, when these are in contact with external environment. Although the XPS analysis in this study does not claim chemical recombination in depth of the investigated layers, such is well possible during sensors' preparation, as indirectly evident from the hereby obtained results and as also reported with respect to C-based substances formation on a Ag substrate. Additionally, impurities within the AgCl layer are also to be denoted to exposure of the sensors to atmospheric conditions within transfer from one to another equipment or setup. These rapid transformations need to be considered with respect to the sensors' preparation and, later on, durability for practical applications.

This section presented and discussed the AgCl layer morphology, microstructure and chemical composition, which vary and depend on the applied current densities within the anodization regimes. Higher current densities result in thicker AgCl layers (as expected), increased complexity (e.g., more than one interface was observed) as well as higher impurities and chemical recombination. These may subsequently influence the potentiometric response of the sensors in the highly alkaline environment of cementitious materials, which is discussed in the next section.

3.3.2   Open Circuit Potential

The OCPs of both steel rods and sensors, as embedded in the cement-paste cylinders, were monitored during 180 days of immersion in simulated pore solution (SPS) with 870 mM chloride concentration. Within the process of SPS penetration, chloride ions also diffuse into the bulk cementitious matrix. Although OCP is a parameter that provides qualitative information only and its interpretation is not straightforward, the alteration in OCP values for both sensors and steel rods follows the changes in their electrochemical state.

3.3.2.1   OCP Development: Steel Rods

Steel, embedded in a cement-based environment is considered to be active (corroding), when the recorded open-circuit potential is more cathodic than -273 mV vs. SCE [30]. The significant shift to active (more cathodic) potential can be also associated with the so-called threshold chloride content. This is because in the event of sufficient chloride ion concentration in the vicinity of the steel surface, a transition from passive to active state would be the dominant process. Figure 3.4 depicts the recorded OCP evolution of the embedded steel in all tested specimens. Should be noted that the specimen designation A, C and D refers to variation in the embedded sensors' preparation only, while all steel rods were identical in preparation, implying identical or at least similar steel surface properties. Having two replicates for A, C and D specimens, the monitored steel rods designation is: A1, A2, C1, C2 and D1, D2 (the steel response in B specimens was almost identical to this in specimens A and for simplicity is not presented and discussed).

Additionally, the time scale in Fig. 3.4 presents the time of conditioning of the cement paste specimens, i.e., $t = 0$ corresponds to the start of exposure to the SPS medium. At $t = 0$, the specimens have already been cured for 30 days and therefore a stable product layer (including passive film) would be present on the steel surface

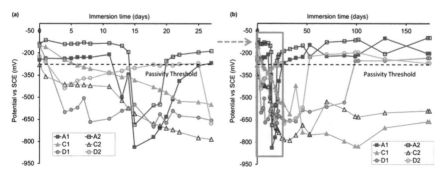

Fig. 3.4 OCP reading of steel rods (**a**) at the initial days and (**b**) over long immersion period into the SPS

in all cases. To that end, variation in the recorded OCP of steel (Fig. 3.4) from 0 to 180 days would depend on the rate of chloride ions penetration and corrosion initiation. Chloride ions penetration, on the other hand, is largely determined by porosity, permeability and diffusivity of the cementitious material. In other words, OCP fluctuation for the steel rods is also related to the properties of the cement paste bulk matrix [9, 31]. This is in addition to the phenomena related to passivity breakdown (corrosion initiation), corrosion propagation and/or repassivation occurrences, all of which are characteristic and well known within the chloride-induced corrosion process in reinforced cement-based materials [9, 31].

As shown in Fig. 3.4a, at the first day of immersion (or 30 days of cement hydration), the potential of all specimens was between −100 and −280 mV vs. SCE, indicating passivity [30]. It is worth noting that various parameters may affect the absolute potential values, without giving information about the actual corrosion state of the steel rods. OCP is not only dependent on the chloride concentration at the steel surface but can be also affected by other factors, such as pH of the pore solution, temperature, microstructure and composition of the concrete-steel interface, or composition and surface finish of the steel [32, 33]. Hence, the interpretation of absolute OCP values may be misleading for the estimation of corrosion activity and can only be used as an indication of the electrochemical state. Therefore, for the evaluation of an ongoing corrosion process on the steel surface, monitoring the OCP development over time can be more useful [34].

Corrosion onset and corrosion propagation can be apparent only from a sustained cathodic drop in the potential. Despite the otherwise identical environmental condition, for identical steel rods, three different trends of OCP development were observed. This accounts for a significantly different electrochemical state within treatment in the SPS environment. While for the two replicates in group A, a cathodic shift in OCP values was observed between 13 and 15 days of treatment, the specimens in groups C and D depict a gradual/abrupt shift to cathodic potentials from the start of immersion (Fig. 3.4a). In the case of specimens A, OCP values around −600 mV can be an indication of corrosion initiation after 2 weeks of exposure to the chloride-containing SPS medium. However, corrosion propagation is uncertain, since these cathodic values were not sustained and OCPs shifted to anodic values, well above the passive/active threshold of −275 mV, after 20–25 days of treatment.

In contrast, the OCPs for specimens in group C showed an almost immediate drop to more cathodic values (Fig. 3.4a). This trend was sustained throughout the experiment: trespassing the passive/active threshold before 5–6 days of treatment, gradual decline between 10 and 50 days (Fig. 3.4a, b) and maintained OCP values around −650 mV until the end of the test (Fig. 3.4b). The OCP values for specimens D depict the largest fluctuation in the course of the experiment, which is in addition to the highest deviation between the replicates (D1 and D2), compared to the behavior of replicates in the other groups (A1 and A2 and C1 and C2). Initially, between 1 and 5 days of immersion, the OCP evolution for D1 and D2 showed development similar to that in groups C1 and C2 – almost immediate cathodic shift (Fig. 3.4a), reaching −650 mV for specimen D1 after 5 days. Between 5 and 25 days of treatment, specimen D1 maintained this cathodic potential, i.e., continued following the pattern of behavior for specimens C (Fig. 3.4a). In contrast, specimen D2 exhibited an anodic shift, starting after 5 days of treatment and followed that line of change (Fig. 3.4a), arriving at more noble potentials (above the active/passive threshold of −275 mV) after 25 days of treatment, followed by a sudden shift to the cathodic values.

What can be concluded from the monitoring of OCP values in all groups at earlier stages of the experiment (Fig. 3.4a) is as follows: OCP fluctuations were expected in the initial and earlier periods of treatment. This is due to the varying speed of external solution penetration (including chloride and/or hydroxide ions) into the bulk cement-based matrix, which determines the different levels of oxygen and relative humidity at the steel/cement paste interface. Next, balancing of these gradients is expected over time, which would be reflected in OCPs stabilization and/or a sustained trend of OCP evolution. This was as observed for the cases of specimens in groups C and D, while an event of a significant cathodic drop in OCP values was observed for specimens A after 2 weeks of treatment. Although, corrosion initiation can be relevant for specimens A after 2 weeks of treatment, corrosion propagation was not sustained.

The cathodic values observed for specimens C and D, immediately after immersion in the SPS medium, were not related to chloride-induced corrosion initiation, but rather denoted to limitations of the electrochemical reactions on the steel surface. A cathodic potential of −600 mV can reflect not only corrosion initiation but can be also an indication of limited oxygen availability for the cathodic reaction on the steel surface in highly alkaline environment, and in submerged condition [2, 3, 35–37]. The low oxygen concentration in submerged condition limits the cathodic current density, and thereby restricts the current flow between anodic and cathodic sites. This results in a more cathodic potential but reduced overall corrosion rate. Meanwhile, the same range of cathodic potentials, but elevated corrosion rates, can be expected in the case of simultaneously occurring chloride-induced corrosion, since this process is well known to be an autocatalytic one [38].

The passivity breakdown for specimens A is followed by a nonsustained corrosion activity. As can be observed, the OCP values for specimens A remained well above the threshold for active state after 20–25 days and towards the end of the experiment (Fig. 3.4b). Similarly, the OCP values for specimens D stabilized in the region of passivity after 50 days for specimen D2 and after 100 days for specimen D1, Fig. 3.4b. Nevertheless, the cathodic OCP values for both D1 and D2 were maintained between 25 and 50(100) days of treatment and can therefore account for corrosion initiation and propagation in these specimens. Attempts for repassivation after this period (after 50 days for D2 and 100 days for D1) and towards the end of the test were reflected by the anodic shift of OCP values. The corrosion kinetics, though, cannot be judged from OCP evolution only and therefore no further conclusions on the corrosion state in specimens D can be derived from Fig. 3.4. Considering the longer time interval of sustained cathodic potentials (>25 days), it can be expected that the corrosion rates in groups C and D were significantly higher than those in group A at the end of the test. Quantification of the electrochemical state is discussed in Sect. 3.3.3.

From all investigated specimens, the steel rods in groups C presented active behavior from the start until the end of the experiment. Although the initial cathodic OCP values can be attributed to the synergetic effect of chloride penetration and limited oxygen availability at the steel/cement paste interface, the sustained cathodic potentials over time reflect not only corrosion initiation but also corrosion propagation. While as abovementioned, the kinetics of corrosion cannot be judged from OCP evolution only, quantification of the electrochemical response (through the hereby employed EIS measurements) was expected to confirm the assumed hypothesis and will be discussed further below in Sect. 3.3.3.

A logic question arises based on the above discussed OCP evolution for specimens A, C and D: if the steel rods were identical and the cement-based specimens identically conditioned, why would the OCP response present such variations? First of all, as aforementioned, OCP can only provide indication for corrosion state, which along with the already discussed limitations of the technique, partly explains the recorded responses. Next to that, a reinforced cement-based system is at a very high level of heterogeneity, meaning that even identically prepared and conditioned samples are never completely alike. To that end, also considering the generally expected and widely reported steel electrochemical behavior in cement-based materials [31], the OCP observations are logic for the time interval of 180 days. A stable trend of behavior for steel in cement-based materials is normally to be achieved for longer periods of conditioning. This is also evident from the more stable behavior of the systems in this study towards the end of the test.

One another note: sustained corrosion propagation is only to be expected when a sufficient chloride concentration is present. Obviously this could be the case for specimens C, but not necessarily for groups A and D. As well known, the critical chloride threshold value determines the time to corrosion initiation but does not (experimentally) define the period over which a substantial corrosion propagation will be at hand. Repassivation is a generally observed phenomenon in alkaline environment, as in this test conditions. Therefore, except the actual chloride concentration, the hydroxide ion concentration and the chloride binding capacity, factors as porosity, permeability and pore interconnectivity of the bulk matrix play a significant role. Their effect is expressed within the corrosion propagation process and can significantly vary, as determined by the intrinsic heterogeneity of a cement-based system. All these properties are not subject to this work, however, they need to be reminded and considered in view of the above and following discussion on different electrochemical behavior of otherwise identical reinforced cement-based systems. Additionally, following the objectives of this work on defining the accuracy of chloride sensors for determination of chloride content in cementitious materials, the chloride threshold for corrosion initiation only is of interest in this study. In order to justify the accuracy of sensor's readings, the derived chloride threshold value (judged from the sensor's response) was coupled to the steel electrochemical response. Similarly to the steel rods, the OCP evolution for the sensors over time was monitored. These results and discussion are presented in what follows.

3.3.2.2   OCP Development: Chloride Sensors

The main concept of defining chloride threshold through the application of Ag/AgCl electrodes (chloride sensors) is based on recording the OCPs for these sensors at the corrosion initiation time for the steel rods. Limitations, source of errors and challenges with regard to employing this approach to reinforced concrete systems were already introduced.

Within the process of chloride penetration into the cement-based specimens, the OCP of the embedded sensors would gradually shift to a more cathodic value. This is as shown in Fig. 3.5, following the electrochemical equilibrium of Ag dissolution and AgCl formation on the surface of the sensor. The OCP evolution is as expected and fundamentally determined by the Nernst equation (Eq. 3.1).

$$E = -0.0214 - 0.05916 \lg \left(a_{cl} - \right) \left(V \; vs \; SCE, 25°C \right) \tag{3.1}$$

In this equation, E is the potential of the sensor and a_{cl}^- is the activity of chloride ions (molarity). The above equation is based on the following electrochemical reaction (Eq. 3.2) on the sensor surface:

$$Ag + Cl^- \leftrightarrow AgCl + e^- \tag{3.2}$$

In highly alkaline environment, as concrete, the chloride sensor will deviate from the Nernstian behavior when the chloride-ion activity (concentration) on the sensor surface is remarkably lower than the activity (concentration) of the interfering hydroxide ions. In this condition, the activity of silver ions in the vicinity of the sensor is determined by an exchange equilibrium, Eq. (3.3):

$$2AgCl + 2OH^- \leftrightarrow Ag_2O + 2Cl^- + H_2O \tag{3.3}$$

The gradual formation of silver oxide/hydroxide compounds shifts the OCP of sensor from about 150 to 99 mV or even more cathodic values, 30 mV [39, 40]. It is

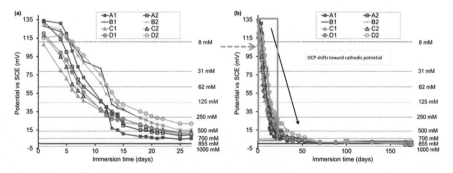

Fig. 3.5 OCP reading of the sensors (**a**) at the initial days and (**b**) over long immersion period into the SPS

known that the silver oxide composition and the hydroxide ions activity in the environment further influence the potential of the already formed silver/silver chloride interface [41]. For example, AgO is a semiconductor with a high conductivity, while Ag_2O has a low conductivity (10^{-8} S/cm). AgO is insoluble in water, while Ag_2O dissolves to the extent of 0.01 g/l [41]. Therefore, it is expected that in the absence of chloride ions, alterations within a complex silver oxide-containing layer can gradually block the pores and impede the faradaic reaction on the sensors' surface. This process is likely to favor the shift of the sensor's potential to more cathodic values (30–99 mV).

It should be bared in mind that the described process for silver oxide formation is in addition to the inherent impurities within the AgCl layer, i.e., oxide-/carbon-based compounds on the sensor surface during the anodization (Fig. 3.3) and before embedding them in the cement paste cylinders. This will possibly result in an initial difference in the OCP values of identical sensors, even prior to their embedment in the cement paste specimens, as was actually observed previously for model medium [21].

In the beginning of immersion, considering the 30-day cured cement paste cylinders, maintained further in sealed condition, the penetration rate of chloride ions from the external solution into the bulk matrix can be considered as minimal. This is justified by the expected, and as recorded, anodic potentials for the sensors, 108–135 mV (Fig. 3.5a), corresponding to a negligible chloride content of lower than 8 mM at the sensor surface (Fig. 3.5a). A note here: if the initial OCP records for the steel rods are considered for that same time interval (Fig. 3.4a), the sensors reading is an additional proof for the reasons behind the already discussed initial cathodic potentials for specimens C and D, i.e., OCP drop was due to reasons, other than chloride ions present in the vicinity of the steel rods.

After 15 days of conditioning, the OCP values of the sensors sharply shifted in the cathodic direction and reached to ~30 mV, which is equivalent to a chloride concentration of ~250 mM (Fig. 3.5a). OCP stabilization was achieved after about 60 days of immersion for all specimens (Fig. 3.5b).

A stable potential is generally accepted to be obtained when the difference in the subsequent measurements is ~1 mV which is considered as a negligible potential difference [17, 42].

Within the process of chloride penetration, the OCP evolution for the different sensors (A, B, C and D, Table 3.1) was similar (Fig. 3.5) and no clear trend for justification of the dependence of OCP readings on the AgCl layer thickness can be found at a first glance. However, a detailed observation of the electrochemical response of the sensors at the initial and final time intervals (Figs. 3.5 and 3.6) and coupling it with the morphological observations (Fig. 3.2) and surface chemical composition (Fig. 3.3) allows a more in-depth approach to find a correlation between the sensor potential response and the preparation procedure.

In the initial stage, the measured OCP of all sensors varied with 14–26 mV, Fig. 3.6a. The OCP of A sensors was 134 mV, for B sensors was 130 mV, while the sensors prepared at regimes C and D showed 108–122 mV and 115–120 mV, respectively (Fig. 3.5a). The distinct OCP difference of the sensors indicates that within

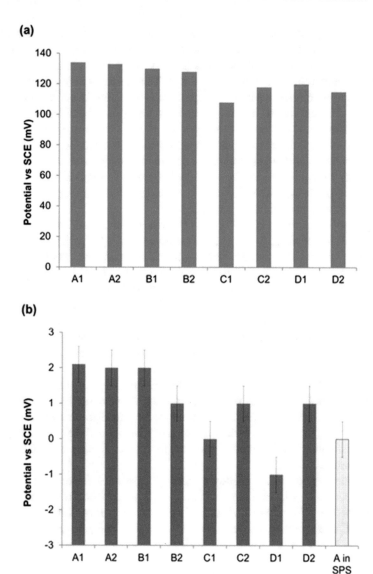

Fig. 3.6 Open circuit potential of sensors in (**a**) the absence of chloride (day 1) and (**b**) after long immersion (day 180)

the process of curing (30 days) in the high pH of the pore solution of cement paste, gradual transformation within the AgCl layers and, therefore, formation of Ag_2O shifted the OCP to more cathodic potentials. The rate of these transformations depends on the morphology, microstructure and composition of the product layer on the sensor's surface.

As previously introduced, and later on commented in Sect. 3.3.2.2, AgCl transforms to AgO, reflected by the measured OCP of the sensors. Obviously, in the case of C and D, the surface of the sensors was a mixture of AgCl and AgO, evident by the initial OCP values around 110–120 mV. The higher heterogeneity – both in terms of morphology and chemical composition – on the surface of C and D, would result in a different rate of the electrochemical reactions within Ag oxidation and reduction later on, if compared to sensors A and B. In the former case, the initial changes of the AgCl layer would account for a delay in detection of relevant chloride concentration with a possibly larger deviation in the sensors' readings. This is because any limitations to electron and ionic transfer (activation or concentration polarization) on the surface of the sensors would result in varying rate of the electrochemical reaction of chloride ions detection.

As previously shown (Fig. 3.2), A and B sensors contain AgCl packed-piled particles on the surface of the Ag substrate. In contrast, the AgCl particles in the sensors C and D are of a smaller size, containing wider inter-grain boundaries (pores) and a multilayer structure (Fig. 3.2). Within the curing time of cement paste cylinders in sealed condition and prior to immersion into the SPS, surface modification of the Ag/AgCl sensors would be as expected. In this period, cement hydration is the determining factor for the chemical and microstructural changes of the bulk matrix and the pore solution of cement paste. Therefore, the development of sensor/cement paste interface would account for the initial chemical/electrochemical reactions on the sensor surface.

Generally, in the highly alkaline, chloride-free pore solution of the cement paste, the AgCl particles would start dissolving in order to maintain the electrochemical equilibrium between the silver and the chloride ions (due to well-known interference of OH$^-$). At the same time, the formation of silver oxide/hydroxide compounds (Eq. 3.3), would gradually modify the sensor surface, shifting the OCP towards values around 110 mV [39].

The wider inter-grain space between the AgCl particles close to the silver substrate in C and D sensors in comparison to A and B sensors accelerates the exchange equilibrium (Eq. 3.3) and the formation of silver oxide layer resulting in initial OCP values for C and D sensors between 108 and 122 mV in comparison to A and B sensors (130–135 mV), (Fig. 3.6a). This difference in initially recorded potentials can be additionally denoted to the impurities and/or chemical transformation in the AgCl layer formed during sensors' handling and within casting after anodization. As previously indicated in Fig. 3.3, the impurities on the AgCl surface account for similar chemical composition of the (composite) outer AgCl layer. However, different inner morphology of the AgCl particles, especially close to the silver substrate, was observed (Fig. 3.2b, h).

On one hand, similar chemical composition of the outer layer and relatively similar OCP of the B and C sensors (108–130 mV), after they have already been embedded for 30 days in the cement matrix, would imply that the inner morphology of the AgCl layer is a less influential factor on the overall electrochemical response. On the other hand, the initial OCP measurements are not really similar, since more anodic OCP of the B sensors (130 mV) was recorded, in comparison to the C sensors (108–118 mV), Fig. 3.6a. A difference of 12–22 mV can be considered significant from both electrochemical viewpoint and in view of sensors' potential records, i.e., chloride content. Therefore, the impact of the inner layer (which is different for A and B sensors) is possibly larger, while surface composition and morphology of the outer layer are contributing, but not solely responsible factors for the different initial response of sensors A and B. Therefore, the morphology, orientation and distribution of the AgCl particles within the inner layer, nearby the Ag substrate, can be considered as factors of high importance for the overall electrochemical response of the sensors.

To this end, and if all types of sensors are considered, the larger the variation in the inner layer, the larger the OCP fluctuation (and more cathodic values) would be expected, irrespective of the surface morphology (outer layer) and composition. This was actually as confirmed by the initially recorded OCP values for sensors C and D (Fig. 3.6a), where OCP values varied between 108 and 120 mV.

The OCPs of the sensors varied not only in the beginning of treatment but also after prolonged conditioning of the cement paste cylinders in the SPS medium. Figure 3.6b presents the final readings (180 days) when a semi-equilibrium was established and a high chloride concentration was already present in the vicinity of the sensors' surface. Within the process of chloride penetration, the OCPs of the sensors shifted to cathodic values and after 180 days stabilized at −1 to 2 mV (Fig. 3.6b). The small variation in the measured OCP (3 mV) indicates negligible interference by hydroxide ions. The stabilized OCPs of A and B sensors were 2 ± 0.5 mV (corresponding to the average chloride concentration of 820 mM), while those for C and D sensors were 0 ± 1 mV (equivalent to the chloride concentration of 850–950 mM). While a difference between 1 mV and 3 mV is considered to be not of a large significance from the view point of global electrochemical performance, such a small difference is obviously significant if the electrochemical response is linked to chloride concentration. The more cathodic potential and larger OCP variation for C and D sensors, in comparison to A and B sensors, are the indication of higher heterogeneity and nonuniform distribution of surface AgCl clusters, resulting in the "detection" of slightly higher chloride concentration. These are as expected, as observed and already previously commented.

Therefore, the sensor preparation regime and the AgCl layer characteristics (e.g., morphology, thickness and composition) affect the potentiometric response of the sensor in the absence and in the presence of chloride ions. Previously it was shown that at 250 mM chloride concentration in the simulated pore solution, the effect of AgCl layer properties on the measured OCP is minimum (1–2 mV variation among the OCPs), owing to a faster rate of electron and ionic exchange mechanisms, which decreases the deviation in the sensors' response [21].

Finally, conditioning of the sensors in cement paste, logically affects the OCP response, if a comparison to sensors' readings in solutions only is considered. An identical replicate of sensor A, as embedded in cement paste, was tested in SPS solution only. The chloride concentration of the SPS solution was 875 mM. The A sensor, embedded in cement paste after 180 days of conditioning reads 2 mV (corresponding to 820 mM chloride concentration), while A sensor, directly immersed into the SPS, shows an OCP value of 0 mV (corresponding to 890 mM chloride concentration) – Fig. 3.6b. The embedded sensor measured lower chloride concentration (~70 mM) in comparison to the sensor directly immersed into the SPS. It is well known that the physical and chemical adsorption of chloride ions on the charged surface of hydrated cement paste products lowers the chloride content in the cement paste pore solution and on the sensor surface, respectively. In other words, a portion of the sensor's surface, in contact with cement paste, might be screened from the direct "access" of chloride ions. The lower chloride content in these locations and the direct contact with hydration products and fine pores moderate the chloride concentration with respect to the total surface of the sensor. This would be reflected by the OCP measurement at that point in time, the sensor A in cement paste ending up with a more anodic OCP (2 mV) if compared to the reading in SPS only (0 mV).

In conclusion, this section discussed the simultaneous monitoring of OCP for both steel and sensors in identical environment. The correlation of results can provide important information on the chloride threshold value for steel corrosion initiation in cement-based materials. However, the accuracy of OCP measurements in view of the time to steel corrosion initiation at the point of derived chloride content (via the sensors readings) needs to be justified. Therefore, in this study EIS, as a nondestructive technique, was employed in order to qualitatively and quantitatively support the above discussed results.

3.3.3 EIS Response of the Steel Rods

Electrochemical impedance spectroscopy (EIS) was employed to evaluate the corrosion state of the steel rods over time, through the determination of polarization resistance (R_p) values and corrosion current densities, respectively.

3.3.3.1 General Considerations

EIS can determine the overall impedance of a system over a certain frequency range. The experimental data are commonly displayed in Nyquist and Bode plots format. The high to medium frequency domain (e.g., from 50 kHz and higher to 10 Hz) denotes to the properties of the bulk cement paste matrix [8, 9]. The low frequency response (10 Hz to 10 mHz) is related to polarization resistance, including transformations (mass transport or redox processes) in the product layer on the steel surface [8, 43–49].

In the present study, the impedance response in the low frequency domain was used to evaluate the variation of corrosion current density (I_{corr}) over time. The apparent I_{corr} was calculated using the well-known Stern-Geary equation, $I_{corr} = B/R_p$, where R_p was obtained from fitting of the EIS response at the low frequency domain

[9] and the constant B was employed at a value of 26 mV (active state) as previously reported [50].

3.3.3.2   Equivalent Circuits

Different equivalent circuits can be used for quantitative analysis of the EIS response of heterogeneous and complex corrosion systems, such as steel in cement-based materials, with equally good fit results. From the mathematical point of view, increasing the complexity of a circuit would decrease the fitting error [51, 52].

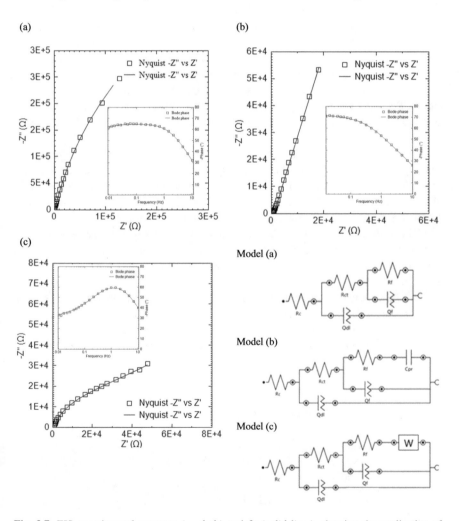

Fig. 3.7 EIS experimental response (*symbols*) and fit (*solid lines*), showing the application of Models (**a–c**) for specimen D1 (*Models a, c*) in the beginning and at the end of immersion, and specimen D2 (*Model b*) within the intermediate period of conditioning

However, the most reliable circuit would be the one with parameters of a clearly defined physical meaning, attributed to the various features of the experimental response [46, 53–55]. The electrochemical impedance spectra were recorded in the frequency range of 50 KHz to 10 mHz, while fitting for evaluation of corrosion parameters was performed in the frequency range of 10 Hz to 10 mHz (i.e., the frequency range, where the response is relevant to the electrochemical reactions on the steel surface [6, 56, 57]).

The electrochemical behavior of the steel rods varied with time in this experiment from 0 to 180 days. This was expressed in a different EIS response, which was not possible to be fitted with only one circuit model. The circuits used in this work are presented in Fig. 3.7. Model (a) in Fig. 3.7a is frequently used in the literature for fitting the EIS spectra for steel in cementitious materials [6, 56, 57]. In this circuit, R_c corresponds to the electrolyte resistance, including the contribution of the bulk matrix (cement paste) ohmic resistance. The first time constant is attributed to the double layer capacitance (Q_{dl}) and the charge transfer resistance (R_{ct}), linked to the electrochemical reactions on the steel surface. Q_{dl} is a constant phase angle element CPE, which is often used in place of an ideal capacitor (C_{dl}) to account for the nonhomogeneity of the system [9]. In some cases, when $n \approx 1$, the CPE is replaced by C_{dl}.

The second time constant corresponds to the redox or mass transport processes in the passive/oxide layer in which R_f and Q_f are the relevant resistance and pseudo-capacitance. At certain time intervals, a constant phase element (Q_{pr}) [58–61] or a Warburg impedance (W) [59, 62] was added in series with R_f (relevant to the low frequency domain). This changed the used circuit from Model (a) to Model (b) or Model (c) (Fig. 3.7b, c). The added capacitive (Q_{pr}) or diffusive (W) elements in the last time constant can be attributed to a diffusion controlled reaction or surface film formation/redistribution due to the ongoing corrosion processes on the steel surface.

Different equivalent circuits were used for one and the same specimen depending on the recorded EIS response, i.e., more than one model was employed in the course of the test in order to account for the changing electrochemical state of the steel electrodes (Fig. 3.8).

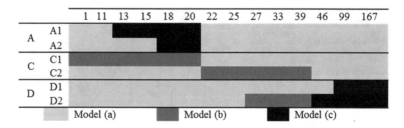

Fig. 3.8 The fitting equivalent circuits for the EIS spectra of the steel rods over the exposure time

The response for specimens A was mostly fitted with Model (a), except for the period of corrosion initiation, where a Warburg impedance (Model c) was added. The time to corrosion initiation, as determined by the EIS response, is discussed further below in the Sect. 3.3.3.3. For C specimens, the employed circuits were Models (a) and (b). For C1 specimen, the EIS spectra were fitted with Model (b) for 20 days from the beginning of the test, after which Model (a) was mainly used. The response for specimens C2 was mostly fitted with Model (a), except for a period between 22 and 33 days when the addition of Q_{pr} to the fitting circuit was necessary, therefore, using Model (b). Similarly to specimens A, the EIS response for specimens C, required a change in the equivalent circuit for data fitting, which was at the time of corrosion initiation.

In D specimens, application of Model (a) resulted in deriving best fit parameters with a lowest fitting error. Model (a) was employed from the start of the experiment until day 46 for D1 and day 27 for D2. Model (c) was used for the rest of the immersion period for D1 specimen, while the short period of using Model (b) in D2 specimen was accompanied by using Model (c) for the rest of immersion period and until the end of exposure. In D1 specimens, corrosion initiated within the time period of implementing Model (a), while similarly to other specimens, the time of corrosion initiation in D2 coincided with the time of changing the circuit.

As previously mentioned, the OCP alteration is not sufficient for justification of the corrosion activity on the steel surface over time. Therefore, determination of corrosion initiation and quantification of steel corrosion in time requires calculation of polarization resistance (R_p) – a parameter linked to the corrosion activity on the steel rods. R_p was calculated by summing up the resistance parameters of EIS fitting circuits (R_{ct} and R_f) and was considered as the global resistance of the embedded steel rods [63, 64].

3.3.3.3   Quantification of EIS Response

As aforementioned, the interpretation of EIS measurements can result in quantitative information for the systems under study. The corrosion current densities can be calculated based on the polarization resistance values. If more accuracy within quantification of electrochemical response is to be achieved, the general approach in electrochemistry is to couple at least two types of electrochemical measurements and correlate their outcomes. In this work, EIS mainly was employed as a nondestructive technique in view of the objective to determine the time for corrosion initiation on steel in correlation to the chloride sensors response. Therefore, although corrosion currents for steel during the tests are presented further below, absolute values are not claimed. The figures and discussion are rather presented in the sense of associating this work to reported literature, where corrosion current density is used as a main parameter of discussion. The derived corrosion current density values in this work are also discussed in comparison to the well-accepted threshold values, i.e., 0.1 to 0.5 $\mu A/cm^2$ [65–68], beyond which steel is considered to be in active (corroding) state.

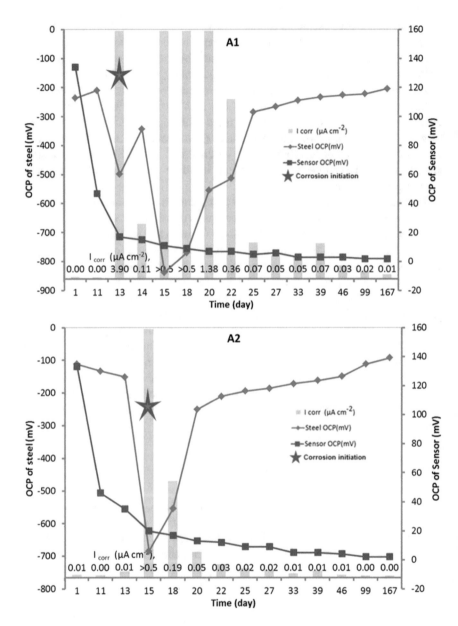

Fig. 3.9 Corrosion current density of steel rods calculated from the EIS fitting circuit parameter (Rp) in addition to the OCP of steel and sensor in A specimens at a certain time over immersion period

Figures 3.9, 3.10 and 3.11 present the corrosion current density of the steel rods over time. The plots simultaneously depict the OCP values of both steel and sensor when embedded together in one and the same cement-paste cylinder. Hence, it was expected that a correlation between the measured OCPs for the sensors, on the one

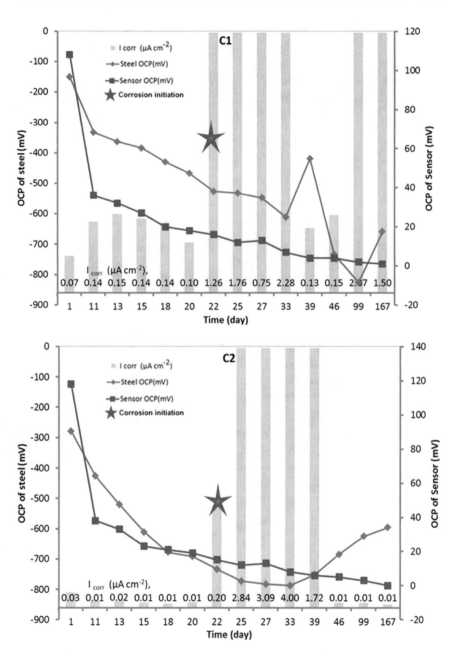

Fig. 3.10 Corrosion current density of steel rods calculated from the EIS fitting circuit parameter (Rp) in addition to the OCP of steel and sensor in C specimens at a certain time over immersion period

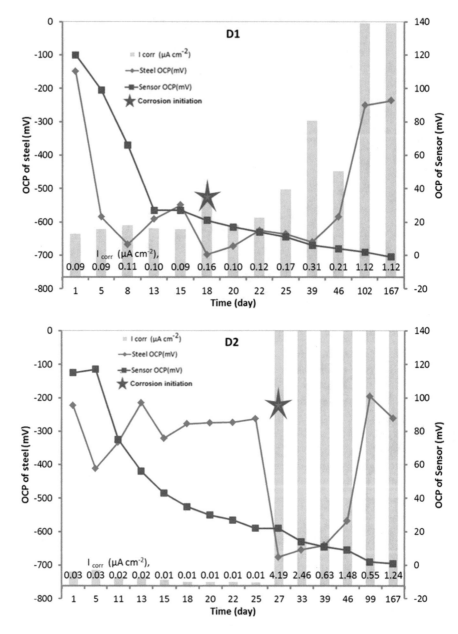

Fig. 3.11 Corrosion current density of steel rods calculated from the EIS fitting circuit parameter (Rp) with respect to the OCP of steel and sensor in D specimens at a certain time over immersion period

hand, and the corrosion current density derived for steel, on the other hand, will provide information about the chloride concentration at the time of corrosion initiation in each specimen. Depassivation is considered to be the starting point of a maintained sharp increase in the corrosion current density, i.e., a sustained increase in the current is to be observed, rather than a singular event of current density increase [31, 69].

It was expected that over 30 days curing of the reinforced cement paste samples in sealed condition, the electrochemical reactions on the steel surface will result in the formation of a passive layer due to the highly alkaline environment. In this condition, low corrosion current density and more anodic OCPs for the steel rods are generally expected to be observed [70–72]. As well known, the critical chloride threshold value determines the time to corrosion initiation but does not (experimentally) define the period over which a substantial corrosion propagation will be at hand. The chloride concentration at the corrosion initiation time can be sufficiently high to initiate local attack, but might not necessarily be able to sustain a stable pit growth. The high alkalinity of the pore solution in cement-based materials, hence the competition between aggressive chloride ions and hydroxide ions (pH = 13.6), governs the processes of pitting and repassivation on steel [70, 72]. Therefore, after corrosion initiation and depending on the local condition on the steel surface, the enhanced formation and stabilization of iron hydroxide/oxide layer (γFe_2O_3, Fe_3O_4, FeO, FeOOH, etc.) either reduces or elevates the corrosion current density. Generally, repassivation is evident by a sharp decrease in the corrosion current density which can be accompanied by simultaneous shift of the OCP to the anodic values.

In A specimens, the low corrosion current density (<0.01 $\mu A/cm^2$) and more anodic potential of the steel rods (> − 273 mV) are the indication of steel passivity during the first 2 weeks after immersion into the SPS (Fig. 3.9). An event of a significant cathodic drop in the OCP values of specimens A was observed after 2 weeks of treatment. This is in agreement with the concurrent increase of corrosion current density. In specimens A, the onset of corrosion is apparent from the simultaneous drop in OCP (−500 to −800 mV) and the rise in corrosion current density (>0.5 $\mu A/cm^2$) after 13 and 15 days of immersion. The simultaneously measured OCP of the sensors reflected the chloride content in the vicinity of the steel surface during the above events. In other words, the response of the sensors at the time of corrosion onset indicates the chloride threshold value for the corrosion initiation of the steel. For specimens A, the chloride threshold values were in the range of 380–440 mM. In the next stage, after 22 days onwards, repassivation was evident from the OCP shift towards more anodic values (> −273 mV) and a simultaneous drop in corrosion current density (<0.1 $\mu A/cm^2$), both remaining in these ranges until the end of the test. What can be concluded from the monitoring of both sensor and steel rods for A specimens is the applicability of the sensor for chloride ion determination as well as a good correlation between the steel and sensors electrochemical responses.

The electrochemical response for specimens C and D was obviously different, if compared to specimens A. In C and D groups, the potential drop of the steel to values more cathodic than −273 mV occurred within a few days after immersion

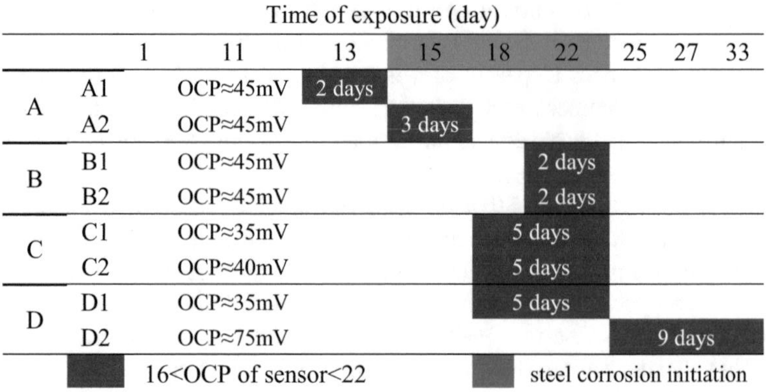

		Time of exposure (day)								
		1	11	13	15	18	22	25	27	33
A	A1		OCP≈45mV	2 days						
	A2		OCP≈45mV		3 days					
B	B1		OCP≈45mV				2 days			
	B2		OCP≈45mV				2 days			
C	C1		OCP≈35mV				5 days			
	C2		OCP≈40mV				5 days			
D	D1		OCP≈35mV				5 days			
	D2		OCP≈75mV						9 days	

16<OCP of sensor<22   steel corrosion initiation

Fig. 3.12 The changes in the OCP of the sensor versus the time to corrosion initiation

(Fig. 3.4a, Sect. 3.3.2.1). At the same time, the sensor's OCP demonstrated a very low chloride concentration (mostly ~8 mM) in the cement paste (Fig. 3.5a, Sect. 3.3.2.2). This amount of chloride ion is much lower than the expected chloride threshold value (45–650 mM) [4]. Meanwhile, the corrosion current densities from EIS measurements were also lower than the threshold value for corrosion initiation ($<$~0.1 µA/cm^2), Figs. 3.10 and 3.11. Therefore, the sustained cathodic OCP value, immediately after immersion in the SPS medium, is not related to the chloride-induced corrosion initiation, but rather denoted to the limitations for the electrochemical reactions on the steel surface (as already discussed). Corrosion initiation was evident by a sharp increase in the corrosion current density between 18 to 27 days of immersion when the chloride concentration, as detected by the sensors' OCP, ranged between 350 and 480 mM. For all specimens C and D, except for specimen C2, the higher than specimens A corrosion activity was maintained for the remaining period of the test.

With respect to the objective of this study, i.e., determination of chloride threshold value, the mathematical calculations show that the time for corrosion initiation of steel is at day 19 ± 5 when the sensor potential is 19 ± 3 mV (equivalent to the chloride threshold value of 400 ± 50 mM), i.e., 70% difference in the time for corrosion initiation and 30% difference among the chloride threshold values are at hand.

The narrow range of chloride threshold values (30% variation) is as expected, since at such high chloride threshold value the influence of the sensor preparation regime on the accuracy of the measurement is negligible. While, the remarkable time difference for steel corrosion initiation (70%) can mainly be attributed to the heterogeneity of the cement paste and the interfacial transition zone between steel and cement paste. Considering the fact that the expected chloride threshold values ranges between 45 mM and 650 mM [4], higher variation among the sensor's measurements can be expected at chloride threshold values lower than 400 mM.

Figure 3.12 presents a correlation chart of the time to steel corrosion initiation (19 ± 5 days) and the sensor response for the relevant chloride threshold (19 ± 3 mV).

As it can be observed, in A and B specimens, the OCP of the sensors change sharply (within 2–3 days) while in C and D specimens the changes take longer (5–9 days). The longer time period needed for C and D sensors to respond can be attributed to the thicker/different inner morphology of AgCl/Ag$_2$O product layer and also to the possible influence of the AgCl layer on the interfacial/microstructural properties at the sensor/cement paste interface. The more complex surface morphology of the C and D sensors, together with varying chemical composition (as previously discussed) in comparison to A and B sensors, may result in a denser packing of cement particles which subsequently influences the chloride content on the surface of the sensors.

3.4   Conclusion

The electrochemical response of the sensors were correlated to the variation of both morphology (including thickness) and composition of AgCl layers, prepared at different anodization regimes. It was shown that the AgCl layer morphology, microstructure and composition vary and depend on the applied current densities within the anodization regimes. Higher current densities result in thicker AgCl layers (as expected), increased complexity (e.g., more than one interface was observed) and higher impurities (formation of compounds other than silver chloride). Among all these variabilities, the silver chloride morphology nearby the Ag substrate is the main parameter influencing the sensor's OCP, i.e., reproducibility and sensitivity, especially at low chloride content, while the surface AgCl layer thickness and impurities are less influential factor.

Despite the identical experimental conditions for this study (e.g., water-to-cement ratio, cement type, exposure conditions, pH of the solution, moisture and chloride content as well as steel rods preparation), there was still a difference in the observed electrochemical responses for the steel and sensor rods. In the beginning of immersion, the sensor preparation regime and therefore, AgCl inner/outer morphology was the main cause for the observed variation of the sensor's OCPs, whereas similar electrochemical response for the steel rods was relevant (OCP > −273 mV and I_{corr} < 0.1 μA/cm^2). Within the process of chloride penetration into the specimens, the inner/outer morphology can moderately influence the arising variability in the measured chloride threshold value of 400 ± 50 mM (30%), the corrosion initiation time of 19 ± 5 days (70%), as well as the overall electrochemical response for both sensor and steel rods. The primary cause for such deviation, especially for the time to corrosion initiation and electrochemical response of steel, appears to be the intrinsic heterogeneity of the adjacent cement paste and variation in the properties of the interfacial transition zone between sensor, steel and cement paste.

The importance of the AgCl layer properties for determination of chloride threshold value should always be considered with respect to the exposure condition and the expected range of chloride threshold value. For example, in splash and tidal zones with lower or largely alternating mean chloride threshold value, sensors with higher sensitivity, reproducibility and accuracy should be used. These are sensors prepared via anodization regimes at low current densities. In contrast, for submerged conditions where higher relative humidity and relatively higher chloride levels are expected, conditions similar to those discussed in this work, the sensor preparation regime is at hand.

References

1. Mehta PK, Monteiro PJM (2006) Concrete: microstructure, properties, and materials. McGraw-Hill, New York
2. Bertolini L, Elsener B, Pedeferri P, Polder R (2004) Corrosion of steel in concrete. Weinheim, Wiley VCH
3. Silva N (2013) Chloride induced corrosion of reinforcement steel in concrete – threshold values and ion distributions at the concrete-steel interface. Dissertation in Civil Engineering (PhD thesis), Goteborg, Chalmers University of Technology
4. Angst U, Vennesland O (2007) Critical chloride content. State of the art- SINTEF RE-PORT
5. Angst U, Elsener B, Larsen CK, Vennesland O (2009) Critical chloride content in reinforced concrete – a review. Cem Concr Res 39(12):1122–1138
6. Andrade C, Keddam M, Novoa XR, Perez MC, Rangel CM, Takenouti H (2001) Electrochemical behaviour of steel rebars in concrete: influence of environmental factors and cement chemistry. Electrochim Acta 46(24–25):3905–3912
7. Song HW, Saraswathy V (2007) Corrosion monitoring of reinforced concrete structures-a review. Int J Electrochem Sci 2:1–28
8. Andradel C, Soler L, Novoa XR (1995) Advances in electrochemical impedance measurements in reinforced concrete. Mater Sci Forum 192–194:843–856
9. Koleva DA (2007) Pulse cathodic protection, an improved cost-effective alternative. PhD thesis, Delft University Press, Delft
10. Poulsen S, Sorensen HE (2012) Chloride threshold value- state of the art. Danish Expert Centre for Infrastructure Constructions
11. Montemor MF, Simoes AMP, Ferreira MGS (2003) Chloride-induced corrosion on reinforcing steel: from the fundamentals to the monitoring techniques. Cem Concr Compos 25(4–5):491
12. Angst U, Vennesland O, Myrdal R (2009) Diffusion potentials as source of error in electrochemical measurements in concrete. Mater Struct 42(3):365–375
13. Elsener B, Zimmermann L, Bohni H (2003) Non-destructive determination of the free chloride content in cement based materials. Mater Corros 54(6):440–446
14. Molina M (1993) Zerstörungsfreie Erfassung der gelösten Chloride im Beton. Diss. ETH, Nr. 10315, ETH Zurich
15. Atkins CP, Scantlebury JD, Nedwell PJ, Blatch SP (1996) Monitoring chloride concentrations in hardened cement pastes using ion selective electrodes. Cem Concr Res 26(2):319–324
16. Atkins CP, Carter MA, Scantlebury JD (2001) Sources of error in using silver/silver chloride electrodes to monitor chloride activity in concrete. Cem Concr Res 31(8):1207–1211
17. Angst U, Elsener B, Larsen CK, Vennesland O (2010) Potentiometric determination of the chloride ion activity in cement based materials. J Appl Electrochem 40(3):561–573
18. Angst UM, Polder R (2014) Spatial variability of chloride in concrete within homogeneously exposed areas. Cem Concr Res 56:40–51
19. Moody GJ, Rigdon LP, Meisenheimer RG, Frazer JW (1981) Selectivity parameters of homogeneous solid –state chloride ion-selective electrodes and the surface, morphology of silver chloride – silver sulphide discs under simulated interference conditions. Analyst 106(1262):547–556

20. Duffo GS, Farina SB, Giordano CM (2009) Characterization of solid embeddable reference electrodes for corrosion monitoring in reinforced concrete structures. Electrochim Acta 54(3):1010–1020
21. Pargar F, Koleva DA, Koenders EAB, van Breugel K (2014) Nondestructive de-termination of chloride ion using Ag/AgCl electrode prepared by electrochemical anodization. In: 13th international conference on durability of building materials and components (DBMC 2014), Sao Paulo
22. Polk BJ, Stelzenmuller A, Mijares G, MacCrehan W, Gaitan M (2006) Ag/AgCl microelectrodes with improved stability for microfluidics. Sensors Actuators 114(1):239–247
23. Shirley DA (1972) High-resolution X-ray photoemission spectrum of the valence bands of gold. Phys Rev B 5(12):4709–4714
24. Scofield JH (1976) Hartree-slater subshell photoionization cross-sections at 1254 and 1487 eV. J Electron Spectrosc Relat Phenom 8(2):129–137
25. Jin X, Lu J, Liu P, Tong H (2003) The electrochemical formation and reduction of a thick AgCl deposition layer on a silver substrate. J Electroanal Chem 542:85–96
26. Bozzini B, Giovannelli G, Mele C (2007) Electrochemical dynamics and structure of the Ag/AgCl interface in chloride-containing aqueous solutions. Surf Coat Technol 201(8):4619–4627
27. Ha H, Payer J (2011) The effect of silver chloride formation on the kinetics of silver dissolution in chloride solution. Electrochim Acta 56(7):2781–2791
28. Lal H, Thirsk HR, Wynne-Jones WFK (1951) A study of the behaviour of polarized electrodes. Part I. The silver/silver halide system. Trans Faraday Soc 47:70–77
29. Fischer T (2009) Materials science for engineering students. Elsevier, San Diego
30. ASTM C 876 (1991) Standard test method for half-cell potentials of uncoated reinforcing steel in concrete
31. Angst UM, Elsener B, Larsen CK, Vennesland O (2011) Chloride induced reinforcement corrosion: electrochemical monitoring of initiation stage and chloride threshold values. Corros Sci 53(4):1451–1464
32. Page CL, Treadaway KWJ (1982) Aspects of the electrochemistry of steel in concrete. Nature 297:109–115
33. Bertolini L, Redaelli E (2009) Depassivation of steel reinforcement in case of pitting corrosion: detection techniques for laboratory studies. Mater Corros 60(8):608–616
34. Elsener B (2001) Half-cell potential mapping to assess repair work on RC structures. Constr Build Mater 15(2–3):133–139
35. Elsener B (2002) Macrocell corrosion of steel in concrete – implications for corrosion monitoring. Cem Concr Compos 24(1):65–72
36. Elsener B, Andrade C, Gulikers J, Polder R, Raupach M (2003) Half-cell potential measurements – potential mapping on reinforced concrete structures, RILEM TC 154-EMC: electrochemical techniques for measuring metallic corrosion, recommendations. Mater Struct 36:461–471
37. Bohni H (2005) Corrosion in reinforced concrete structures. CRC, New York
38. Davis JR (2000) Corrosion: understanding the basics. ASM International, Materials Park
39. Svegl F, Kalcher K, Grosse-Eschedor YJ, Balonis M, Bobrowski A (2006) Detection of chlorides in pore water of cement based materials by potentiometric sensors. J Rare Metal Mater Eng 35(3):232–237
40. Climent-Llorca MA, Viqueira-Perez E, Lopez-Atalaya MM (1996) Embeddable Ag/AgCl sensors for in situ monitoring chloride contents in concrete. Cem Concr Res 26:1157–1161
41. Austin LG, Lerner H (1965) Review of fundamental investigations of silver oxide electrodes. Pennsylvania State University, University Park
42. Angst U, Vennesland O (2009) Detecting critical chloride content in concrete using embedded ion selective electrodes–effect of liquid junction and membrane potentials. Mater Corros 60(8):638–643
43. Ford SJ, Shane JD, Mason TO (1998) Assignment of features in impedance spectra of the cement-paste/steel system. Cem Concr Res 28(12):1737–1751

44. Poupard O, Ait-Mokhtar A, Dumargue P (2004) Corrosion by chlorides in reinforced concrete: determination of chloride concentration threshold by impedance spectroscopy. Cem Concr Res 34(6):991–1000
45. Trabanelli G, Monticelli C, Grassi V, Frignani A (2005) Electrochemical study on inhibitors of rebar corrosion in carbonated concrete. Cem Concr Res 35(9):1804–1813
46. Subramaniam KV, Bi M (2009) Investigation of the local response of the steel–concrete interface for corrosion measurement. Corros Sci 51(9):1976–1984
47. Riberto DV, Souza CAC, Abrantes JCC (2015) Use of electrochemical impedance spectroscopy (EIS) to monitoring the corrosion of reinforced concrete. IBRACON Struct Mater J 8(4):529–546
48. Serdar M, Meral C, Kunz M, Bjegovic D, Wenk HR, Monteiro PJM (2015) Spatial distribution of crystalline corrosion products formed during corrosion of stainless steel in concrete. Cem Concr Res 71:93–105
49. Tang F, Chen G, Brow RK (2016) Chloride-induced corrosion mechanism and rate of enamel- and epoxy-coated deformed steel bars embedded in mortar. Cem Concr Res 82:58–73
50. Andrade C, Gonzalez JA (1978) Quantitative measurements of corrosion rate of reinforcing steels embedded in concrete using polarization resistance measurements. Mater Corros 29(8):515–519
51. Thompson NG, Lawson KM (1991) An electrochemical method for detecting ongoing corrosion of steel in a concrete structure with CP applied. Strategic Highway Research Program, National Research Council, Washington, DC
52. Feliu V, Gonzalez JA, Andrade C, Feliu S (1998) Equivalent circuit for modelling the steel-concrete interface. I. Experimental Evidence and theoretical predictions. Corros Sci 40(6):975–993
53. Andrade C, Soler L, Alonso C, Novoa XR (1995) Importance of geometrical considerations in the measurement of corrosion by means of A.C. impedance. Corros Sci 37(12):2013–2023
54. Gu P, Beaudoin JJ (1998) Estimation of steel corrosion rate in reinforced concrete by means of equivalent circuit fittings of impedance spectra. Adv Cem Res 10(2):43–56
55. Koleva DA, Boshkov N, van Breugel K, de Wit JHW (2011) Steel corrosion resistance in model solutions, containing waste materials. Electrochim Acta 58(30):628–646
56. Joiret S, Keddam M, Novoa XRN, Perez MCP, Rangel C, Takenouti H (2002) Use of EIS, ring-disk electrode, EQCM and Raman spectroscopy to study the film of oxides formed on iron in 1M NaOH. Cem Concr Compos 24(1):7–15
57. Abreu CM, Cristobal MJ, Losada R, Novoa XR, Pena G, Perez MC (2006) Long-term behaviour of AISI 304L passive layer in chloride containing medium. Electrochim Acta 51(8–9):1881–1890
58. Andrade C, Blanco VM, Collazo A, Keddam M, Novoa XR, Takenouti H (1999) Cement paste hardening process studied by impedance spectroscopy. Electrochim Acta 44(24):4313–4318
59. Wei J, Fu XX, Dong JH, Ke W (2012) Corrosion evolution of reinforcing steel in concrete under dry/wet cyclic conditions contaminated with chloride. J Mater Sci Technol 28(10):905–912
60. Morozov Y, Castela AS, Dias APS, Montemor MF (2013) Chloride-induced corrosion behavior of reinforcing steel in spent fluid cracking catalyst modified mortars. Cem Concr Res 47(4):1–7
61. Duarte RG, Castela AS, Neves R, Freirea L, Montemor MF (2014) Corrosion behavior of stainless steel rebars embedded in concrete: an electrochemical impedance spectroscopy study. Electrochim Acta 124:218–224
62. Zhao B, Li JH, Hu R-G, Du RG, Lin CJ (2007) Study on the corrosion behavior of reinforcing steel in cement mortar by electrochemical noise measurements. Electrochim Acta 52(12):3976–3984
63. Jamil HE, Montemor MF, Boulif R, Shriri A, Ferreira MGS (2004) An electrochemical and analytical approach to the inhibition mechanism of an amino-alcohol-based corrosion inhibitor for reinforced concrete. Electrochim Acta 49(5):836

64. Zheng H, Li W, Ma F, Kong Q (2014) The performance of a surface-applied corrosion inhibitor for the carbon steel in saturated Ca(OH)2 solutions. Cem Concr Res 55:102–108
65. Castellote M, Andrade C, Alonso C (2002) Accelerated simultaneous determination of the chloride depassivation threshold and of the non-stationary diffusion coefficient values. Corros Sci 44(11):2409–2424
66. McCarthy MJ, Giannakou A, Jones MR (2004) Comparative performance of chloride attenuating and corrosion inhibiting systems for reinforced concrete. Mater Struct 37(10):671–679
67. Montes P, Bremner TW, Lister DH (2004) Influence of calcium nitrite inhibitor and crack width on corrosion of steel in high performance concrete subjected to a simulated marine environment. Cem Concr Compos 26:243–253
68. Cusson D, Qian S, Chagnon N, Baldock B (2008) Corrosion-inhibiting systems for durable concrete bridges. I: five-year field performance evaluation. J Mater Civil Eng 20(20):20–28
69. COST 52 (2002) Final report, Corrosion of steel in reinforced concrete structures, Luxembourg
70. Li L, Sague AA (2001) Chloride corrosion threshold of reinforcing steel in alkaline solutions-open-circuit immersion tests. Corrosion 57(1):19–28
71. Gouda VK (1970) Corrosion and corrosion inhibition of reinforcing steel: I. Immersed in alkaline solutions. Br Corros J 5(5):198–203
72. Moreno M, Morris W, Alvarez MG, Duffo GS (2004) Corrosion of reinforcing steel in simulated concrete pore solutions: effect of carbonation and chloride content. Corros Sci 46(11):2681–2699

Chapter 4
The Influence of Stray Current on the Maturity Level of Cement-Based Materials

A. Susanto, Dessi A. Koleva, and Klaas van Breugel

Abstract This work reports on the influence of stray current on the development of mechanical and electrical properties of mortar specimens in sealed and water-submerged conditions. In the absence of concentration gradients with external environment (sealed conditions) or in their presence (submerged conditions), compressive strength and electrical resistivity change due to: cement hydration alone; cement hydration, affected by diffusion (including leaching-out); or cement hydration, simultaneously influenced by diffusion and migration. The results are compared to equally conditioned control specimens, where stray current was not involved.

In view of material properties development over time, the ageing factor in relevant exposure conditions is addressed, considering reported approaches for its determination. Through implementing existing methodology and based on experimentally derived electrical resistivity values, the ageing factor for sealed conditions was determined. The apparent diffusion coefficients were calculated based on ageing factors and reported relationships, reflecting the effect of stray current on matrix diffusivity.

Two levels of electrical current density, 100 mA/m^2 and 1 A/m^2, were employed as a simulation of stray current to 28 days-cured mortar specimens with water-to-cement ratio of 0.5 and 0.35. For the time interval of these tests of ca. 110 days, the experimental results show the positive effect of stray current on mortar specimens in sealed conditions and the negative effect for water-submerged conditions.

For sealed specimens, increase of compressive strength and electrical resistivity were recorded, more pronounced for the higher current density level of 1 A/m^2.

A. Susanto (✉) • D.A. Koleva • K. van Breugel
Faculty of Civil Engineering and Geosciences, Delft University of Technology,
Section of Materials and Environment, Stevinweg 1, 2628 CN Delft, The Netherlands
e-mail: a.susanto@tudelft.nl

© Springer International Publishing AG 2017
L.E. Rendon Diaz Miron, D.A. Koleva (eds.), *Concrete Durability*,
DOI 10.1007/978-3-319-55463-1_4

This effect was irrespective of w/c ratio. Increased electrical resistivity and superior performance overall, would determine improved material properties in terms of reduced permeability and diffusivity of the matrix. The results show that for sealed specimens in stray current conditions, the apparent diffusion coefficient was reduced, the effect being more pronounced for the higher current density level of $1 A/m^2$ and (logically) for the lower w/c ratio of 0.35.

In contrast, for water-submerged mortar, a reversed trend of material behavior was observed i.e. reduced mechanical and electrical properties were recorded to be resulting from stray current flow.

Defining a threshold for a positive or negative stray current effect was not possible to be determined from this work. Higher current density levels in varying external environment are necessary to be studied, in order to potentially define such threshold. However, the results clearly show that stray current affects the development of material properties of cement-based materials, an aspect that is rarely considered in the current practice.

Keywords Aging factor • Cement-based materials • Electrical resistivity • Stray current

4.1 General Introduction

The aspects of concrete structures' durability have always been a main concern for the engineering practice. Numerous methods have been proposed to predict service life and reliably foresee structures' performance within their designed life span. One of the important parameters, employed to predict service life, is the so-called "aging factor." The aging factor depends on variables such as the type of cement used and the mix proportion, as well as environmental exposure conditions. All these are further commonly used to predict concrete diffusivity, which is of great significance with respect to durability-related properties of concrete and cement-based materials overall.

Several approaches are known to determine the aging factor of cement-based materials and link these outcomes with concrete diffusivity. Among these, common techniques include natural diffusion tests and accelerated tests, involving the application of electrical current (or voltage) [1–6]. Numerous studies report on different methods for determining concrete diffusivity and chloride transport in cement-based materials [1–10]. Some of these focus on the relationship between nonsteady state diffusion, nonsteady state migration, and steady-state migration tests, referring to test setup as specified in NT Build 443, NT Build 492, and NT Build 355 [5, 7–9]. Other approaches link electrical properties and maturity levels of cement-based materials with diffusivity and aging factors, respectively [11–15].

Determination of water and chloride ions transport in porous medium as concrete, properties as diffusivity, electrical resistivity, maturity, etc., are directly linked to durability, service life, and the aging factor, respectively, of cement-based materials. The transport of fluids and ionic species occurs according to four transport mechanisms such as diffusion due to a concentration gradient, migration due to an electrical potential gradient, permeation due to a pressure gradient, and absorption due to capillary action [6, 16, 17]. These mechanisms are strongly related to and determine the microstructural properties of cement-based materials. Microstructural properties, in turn, determine the global performance of cement-based structures. However, cement-based microstructure development, as governed by cement hydration, would be affected by the abovementioned transport mechanisms, which will progress with various rates if migration i.e. electrical current is involved. Therefore, it is important to clearly differentiate the effect of migration in conditions where ion and water transport are limited to the internal pore water only, from these, where transport mechanisms are largely depended on environmental exposure and conditions. In this work, global performance and properties of sealed and water submerged mortar specimens (compressive strength and electrical resistivity) were derived in rest (control) conditions and in conditions of applied electrical field (stray current). The objective was to clearly differentiate material properties development due to cement hydration only from properties development, additionally affected by altered diffusion and migration, as within stray current conditions. The technical background in the following section summarises the relevant state-of-the-art on main points for material properties evaluation, together with analytical models for deriving parameters of interest (e.g. ageing factor and apparent diffusion coefficients), linking these to the results and discussion in this work.

4.2   Technical Background

4.2.1   Transport Mechanisms and Diffusion Coefficients

4.2.1.1   Diffusion Tests

The natural diffusion test is commonly used to simulate the natural process of chloride transport in concrete as a porous medium. The method is based on diffusion cell tests and immersion tests, employing chloride concentration simulating sea water, i.e., approximately 3.5 wt% [6]. The immersion tests are essentially an immersion of the specimens in a solution, containing a constant chloride concentration. The chloride penetration over time is recorded by grinding the specimen and analyzing the chloride concentration in a direction from the exposed surface towards the bulk material. The result is obtaining a chloride profile after a certain time of immersion. Next, the apparent chloride diffusion coefficient is determined by curve-fitting of the measured chloride profile to the error function in the analytical solution of Flick's second law (Eq. 4.1). The results strongly depend on the immersion period and the chloride concentration in the external bulk solution [6]. In contrast, for the diffusion cell test following NT BUILD 443 [7], where relevant considerations and experimental setup are described in detail, a water-saturated concrete specimen is exposed on one plane surface only to sodium chloride solution (chloride concentration in the range of 3–20%). The chloride content of the cement-based matrix at certain exposure time-intervals is determined within thin layers, ground off in parallel to the exposed face of the specimen. According to NT BUILD 443, the diffusion coefficient can be calculated as follows:

$$C(x,t) = C_s - (C_s - C_i).\mathrm{erf}\left(\frac{x}{\sqrt{4.D_e.t}}\right) \qquad (4.1)$$

where $C(x,t)$ is the chloride concentration, measured at the depth x at exposure time t (mass %), C_s is the boundary condition at the exposed surface (mass %), C_i is the initial chloride concentration measured on the concrete slice boundary condition at the exposed surface (mass %), x is the depth below the exposed surface (m), D_e is the effective chloride diffusion coefficient (m^2/s), t is the exposure time, and erf is the standard error function that can be expressed as below:

$$\mathrm{erf}(z) = \frac{2}{\sqrt{\pi}}\int_z^0 \exp(-y^2)dy \qquad (4.2)$$

Except the above considerations on determining chloride diffusion coefficients and concrete diffusivity, respectively, the application of chloride diffusion tests has also been reported as an aging factor determination approach [1, 10, 18]. The main aspects of this approach are as follows: Fick's second law (Eq. 4.1) is commonly used to calculate chloride profiles at a certain depth and time. In order to predict service life of concrete structures, the effective chloride diffusion coefficient (D_e) in Eq. (4.1) is modified and governed by the following equation [10, 19]:

$$D(t) = D_0\left(\frac{t}{t_0}\right)^{-n} \qquad (4.3)$$

where D_0 is the chloride diffusion coefficient at the reference time t_0 (usually 28 days age), n is the aging factor ($0 \leq n \leq 1$), and t is the age of concrete.

4.2.1.2 Migration Tests

Since natural diffusion is a very slow process, accelerated diffusion test methods are also used, e.g., the rapid chloride migration (RCM) test where chloride ions migration, as a predominant transport mechanism due to the applied voltage in the RCM cell, is in parallel with diffusion of chloride ions (details of the RCM tests are as described in [1, 10] and not subject to elaboration here). Similarly to diffusion tests, the outcomes for concrete diffusivity and diffusion coefficients derived from migration tests can be employed to determine the aging factor for the tested concrete specimens. This approach employs the Nernst-Einstein equation, according to which the chloride diffusion coefficient for a porous material (as concrete) is inversely proportional to the electrical resistivity of this material [20]:

$$D_{cl} = \frac{K}{\rho}, \text{ with } K = \frac{R.T}{z^2.F^2} \frac{t_{cl}}{\gamma_i.c_i} \tag{4.4}$$

where D_{cl} is the diffusivity for chloride ions, R is the gas constant, T is the absolute temperature, z is the ionic valence, F is the Faraday constant, t_i is the transfer number of the chloride ions, γ_i is the activity coefficient for chloride ions, c_i is the chloride ions concentration in the pore water, and ρ is the electrical resistivity.

Determination of the aging factor via records of electrical resistivity development of cement-based materials (discussed further below), using the above approach has also been reported [17–19]. Following Eqs. (4.2) and (4.3), mathematical expression similar to Eq. (4.3) can be obtained, but already introducing the aging factor "q" as follows [10, 21]:

$$\rho(t) = \rho_o \left(\frac{t}{t_o} \right)^{+q} \tag{4.5}$$

$$\text{with } q = 0.798n - 0.0072 \approx 0.8n \tag{4.6}$$

The present state-of-the-art reports aging factor values derived by RCM in the range of 0.178 to 0.65 for CEM I 42.5 [22–26]. This range is in line with values determined by diffusion tests in the range of 0.218 to 0.701 [21, 23]. The range of these values are more or less in the order of aging factor values, as determined through other methods, e.g., via recorded electrical resistivity, a commonly used approach which is briefly introduced in what follows.

More recently, the determination of the aging factor of cement-based materials, based on derived electrical resistivity values (Eq. 4.4 above), was also suggested, reporting values in the range of 0.09 to 0.35 for CEM I 42.5 [11–13]. These are also more or less in line with values, recorded through diffusion or migration tests. Hence, determination of the aging factor based on electrical resistivity values is another suitable approach [11, 14]. However, the accuracy of resistivity determination would depend on the chosen method. This will affect the derived aging factor, respectively.

A brief summary of the most frequently used methods for measuring electrical resistivity of cement-based materials, as well as the approach to maturity levels' determination, is presented in what follows.

Fig. 4.1 Electrical
resistivity measurement by
two probes method

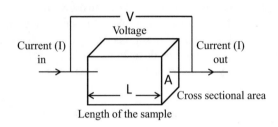

4.2.2   Electrical Properties and Maturity Levels

4.2.2.1   Electrical Resistivity

Monitoring electrical resistivity, a parameter directly linked to maturity of cement-based materials and also related to diffusivity, has been suggested as a convenient and nondestructive technique to assess concrete durability. Electrical resistivity of cement-based materials can be measured following various setups, review of which is not subject to this chapter and can be found described in detail in [27, 28, 29]. The general approach is to record resistance values, by applying an alternating current through a cross section of the specimens, and the resulting output of voltage (or vice versa). Basically the resistance R of a specimen is determined by Ohm's law, $R = V/I$ where V is the electrical potential (in voltage) and I is the applied current (in Ampere).

4.2.2.2   Methods for Deriving Electrical Resistivity

The manner of measuring resistivity values can be different, e.g., two probes measurements, four probes measurements (Wenner configuration), involving the rebar network as one electrode, etc., are generally used.

The two-probe measurement is the simplest method of measuring electrical resistivity – a schematic illustration is presented in Fig. 4.1. In this method, alternating current is applied to the specimens via metal plates of surface area A, equal to the sides (cross-sections) of the specimens (Fig. 4.1). The electrical potential across the specimens is measured and the electrical resistivity is calculated using the following equation:

$$\rho = \frac{RA}{l} \tag{4.7}$$

where ρ is the electrical resistivity of the sample (in Ohm.m), R is the resistance (in Ohm), A is the cross-section of the sample (in m^2), and l is the length of the sample (in m).

The two-probe method is largely used for lab conditions, due to simplicity for lab tests, but also because of the possibility to precisely define geometrical constants and minimize other contributing factors within measurements. For example, the two

Fig. 4.2 Four electrodes configuration according to Wenner [30]

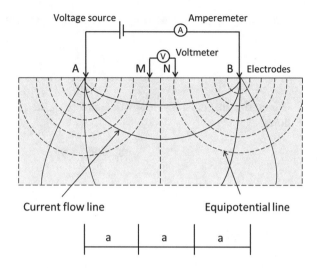

probe method can be executed at constant relative humidity or in sealed conditions, can overcome gradients of humidity or foreign resistance contribution (due to varying concrete cover thickness for example), and can be employed on specimens, which are further used for other tests (e.g., compressive strength).

The second method to measure electrical resistivity of concrete is the four probes measurement, which is developed from geophysical surveying to overcome many of the difficulties/limitation of the two-probe/point method. This method has been used for determining soil resistivity [30] and was applied to concrete structures by Stratfull [31], Naish et al. [32], and Millard [33]. The four-probe measurement is based on the so-called Wenner configuration. The method is largely applied for field tests of concrete and reinforced concrete structures, by using a portable (Wenner probe) device.

According to Wenner, the four electrodes are aligned with the same distance as illustrated in Fig. 4.2. The electrical current is "injected" via the outer electrodes (i.e., electrodes A and B), whereas the electrical potential is measured between the inner electrodes (i.e., electrodes M and N).

The electrical potentials at any nearby surface point are affected by the current flow at both current electrodes A and B (I_A and I_B). The electrical potential due to I_A at point M for example (i.e., V_{MA} and V_{MB}) can be presented as follow [30, 34]:

$$V_M = V_{MA} + V_{MB} \tag{4.8}$$

$$\text{with } V_{MA} = \frac{I\rho}{2\pi r_{MA}} \text{ and } V_{MB} = -\frac{I\rho}{2\pi r_{MB}}. \tag{4.9}$$

where V_M is the electrical potential at point M (in Volt), V_{MA} is the electrical potential between point M and point A, V_{MB} is the electrical potential between point M and point B, I is the electrical current (in Ampere), ρ is the electrical resistivity (Ohm. m), r_{MA} is the distance between point M and point A, and r_{MB} is the distance between point M and point B.

Following Eqs. (4.8) and (4.9), the resulting voltage at point M is:

$$V_{\mathrm{M}} = \frac{I\rho}{2\pi}\left(\frac{1}{r_{\mathrm{MA}}} - \frac{1}{r_{\mathrm{MB}}}\right) \tag{4.10}$$

Similarly, the electrical potential due to I_{B} at point N can be obtained as follows:

$$V_{\mathrm{N}} = V_{\mathrm{NA}} + V_{\mathrm{NB}} \tag{4.11}$$

$$\text{with } V_{\mathrm{NA}} = \frac{I\rho}{2\pi r_{\mathrm{NA}}} \text{ and } V_{\mathrm{NB}} = -\frac{I\rho}{2\pi r_{\mathrm{NB}}} \tag{4.12}$$

finally at point N:

$$V_{\mathrm{N}} = \frac{I\rho}{2\pi}\left(\frac{1}{r_{\mathrm{NA}}} - \frac{1}{r_{\mathrm{NB}}}\right) \tag{4.13}$$

where V_{N} is the electrical potential at point N (Volt), V_{NA} is the electrical potential between point N and point A, V_{NB} is the electrical potential between point N and point B, I is the electrical current (in Ampere), ρ is the electrical resistivity (Ohm. m), r_{NA} is the distance between point M and point A, and r_{NB} is the distance between point M and point B.

It should be noted that the opposite sign of electrical potential (V_{MA} positive, V_{MB} negative and V_{NA} positive, V_{NB} negative) is attributed to the reversed direction of electrical current flow. The difference of electrical potential between V_{N} and V_{M} can be calculated using the following equation:

$$\Delta V = V_{\mathrm{M}} - V_{\mathrm{N}} = \frac{I\rho}{2\pi}\left\{\left(\frac{1}{r_{\mathrm{MA}}} - \frac{1}{r_{\mathrm{MB}}}\right) - \left(\frac{1}{r_{\mathrm{NA}}} - \frac{1}{r_{\mathrm{NB}}}\right)\right\} \tag{4.14}$$

where ΔV is the electrical potential difference between two points, i.e., V_{M} and V_{N}. Next, the apparent resistivity (ρ) is obtained by rearranging Eq. (4.14) as follows:

$$\rho = \frac{2\pi\Delta V}{I}p \tag{4.15}$$

$$\text{with } p = \frac{1}{\left\{\left(\dfrac{1}{r_{\mathrm{MA}}} - \dfrac{1}{r_{\mathrm{MB}}}\right) - \left(\dfrac{1}{r_{\mathrm{NA}}} - \dfrac{1}{r_{\mathrm{NB}}}\right)\right\}} \tag{4.16}$$

where the value of p depends on the electrode geometry.

Fig. 4.3 Setup of one
electrode (disc)
measurement [35]

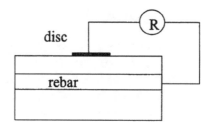

For the Wenner configuration, i.e., $r_{MA} = r_{NB} = a$ and $r_{MB} = r_{NA} = 2a$, the value of p is equal to a. Finally, the apparent resistivity in Eq. (4.15) becomes [30, 34]:

$$\rho = 2\pi a \frac{\Delta V}{I} \tag{4.17}$$

By applying Ohm's Law, Eq. (4.17) can be rewritten as:

$$\rho = 2\pi a R \tag{4.18}$$

where ρ is the electrical resistivity (in Ohm. m), a is the distance between the closest two electrodes (in m), and R is the electrical resistance (in Ohm).

The third method for measuring electrical resistivity of concrete is by involving the rebar network as one electrode [35]. This method is performed by placing a metal electrode on the concrete surface and measuring the resistance between this electrode and the reinforcement as shown in the Fig. 4.3. Basically, this method can be categorized as two-electrodes/probes type measurement. The electrical resistivity of concrete is calculated using the following equation:

$$\rho = k * R(\text{disc} - \text{bar}) \tag{4.19}$$

where ρ is electrical resistivity (in Ohm m), k is the cell constant, and R(disc-bar) is the resistance between the disc electrode and the steel bar (in Ohm).

The cell constant k is quite a complex parameter because it depends on the disc size, the concrete cover, the rebar spacing, and the rebar diameter. According to Feliu et al. [36], for disc sizes smaller than the distance to a large electrode (the rebar system), the electrical resistivity of concrete is expressed by:

$$\rho = 2 * a * R(\text{disc} - \text{bar}) \tag{4.20}$$

where a is the diameter of the disc (in m).

4.2.2.3 Maturity Method

Another method that can correlate aging phenomena and durability of cement-based materials and can be, therefore, applied for aging factors determination, is the "maturity method" [15]. The "maturity" method is a commonly used approach to predict concrete strength development, based on the temperature history of cement hydration. In general, concrete strength development is estimated by using the relationship between the maturity index and strength. The ASTM C 1074 elaborates on the procedure of this standard practice, where the maturity index can be expressed either as a temperature-time factor, using the Nurse-Saul equation, Eq. (4.21) or as the specific temperature at an equivalent age, using the Arrhenius equation, Eq. (4.22),

$$M = \sum_{t}^{0} (T - T_0) \Delta t \tag{4.21}$$

where M is the maturity index (in °C-hours or °C-days), T is the average concrete temperature (in °C), T_0 is the datum temperature (usually taken to be -10 °C), t is the elapsed time (in hours or days), and ΔT is the time interval (in hours or days).

$$t_e(T_r) = \sum_{t}^{0} e^{\frac{E}{R}\left(\frac{1}{273+T_r} - \frac{1}{273+T_c}\right)} \Delta t \tag{4.22}$$

where t_e is the equivalent age at the reference temperature (in hours), E is the apparent activation energy (in J/mol), R is the universal gas constant (in 8.314 J/mol-K), T_r is the absolute reference temperature (in Kelvin), T_c is the average concrete temperature during the time interval Δt (in Kelvin), and Δt is the chronological time interval between temperature measurements (in hours).

According to [15], concrete specimens of the same mix design and at the same maturity level have approximately the same strength, irrespective of the variance in relevant temperature and time (or their combination) in order to make up that maturity. In cement-based materials, strength increases with the progress of cement hydration. The amount of hydrated cement is determined by the duration of curing and the temperature level.

4.2.3 The Contribution of This Work

In view of the introduced technical background and state-of-the-art, this section outlines the contribution of the present work. In altered environmental conditions e.g. when electrical current flows through cement-based materials, the temperature development within cement hydration will change. A temperature increase would be expected as a consequence from ions and water migration, in addition to diffusion phenomena at interfaces e.g. pore wall/pore water. Hence, altered water and ion transport in the cement-based bulk matrix and (re)distribution of hydration products will be at hand. If compared to control conditions, electrical current flow initially leads to accelerated cement hydration and increased strength due to densification of the bulk matrix [37–39]. Various experiments on electrical curing and application of the so-called "maturity method" approach [15] for cement-based materials have been performed and reported [40–44]. The major outcome from these studies is mainly related to prediction of long-term mechanical properties (e.g. strength) and thermal conditions (e.g. heat development) from short-term tests and derived sets of experimental data.

Considering the above relationships, the objective of this work was to evaluate the effect of stray current on the mechanical and electrical properties of 28-days cured mortar in two very distinct exposure conditions (sealed and water-submerged). The motivation for these studies was in view of two main points. Firstly, stray current effects on reinforced concrete structures are mainly considered and reported with respect to steel corrosion, although reported loss of bond strength, coarsening of the pore network, etc. are among effects related to the cement-based material itself [27, 39, 45]. In other words, steel corrosion is well recognised as a consequence of stray current, while the mechanical and electrical properties of the cement-based bulk material are not considered to be subject to change and rather neglected in the present state-of-the-art. Therefore this aspect was studied in this work.

Secondly, if environmental conditions are taken into account, the presence of concentration gradients (e.g. as in underground structures or water-submerged conditions) or the absence of such (e.g. the bulk concrete in large structures, as in sealed conditions) would determine different level of material development in conditions of stray current. In these conditions, ion and water transport are determined by both diffusion and migration. However, in the former case (concentration gradients present), leaching-out would also be relevant and depend on the current density levels to a large extent. In the latter case (no concentration gradient) the effect of stray current would be expressed only in enhanced internal ion and water transport, potentially resulting in altered cement hydration due to migration-controlled phenomena. Therefore, evaluation of material properties in sealed conditions, will explicitly reflect the effect of stray current, when no ion or water exchange with external medium are relevant. In contrast, the development of material properties in water-submerged conditions would reflect the effect of stray current on both ion and water diffusion and migration, together with the contribution of leaching-out effects.

In view of the above and with regard long-term behaviour of cement-based materials in conditions of stray current, the ageing factors and apparent diffusion coefficients were also derived for sealed conditions, based on experimentally recorded electrical resistivity and implementing the above discussed and reported relationships. These were linked to the effect of stray current, rather than derived as absolute values or for the purpose of service life predictions. The outcomes were compared to the relevant control cases and used to evaluate the overall performance of cement-based materials, when ion and water migration contribute to diffusion-controlled transport mechanisms.

4.3   Experimental Materials and Methods

4.3.1   Materials

Mortar cubes of 40 mm × 40 mm × 40 mm (Fig. 4.1) were cast, using OPC CEM I
42.5 N with w/c ratio of 0.5 and 0.35. The cement-to-sand ratio used was 1:3. The
chemical composition (in wt. %) of CEM I 42.5 N (ENCI, NL) is as follows: 63.9%
CaO; 20.6% SiO_2; 5.01% Al_2O_3; 3.25% Fe_2O_3; 2.68% SO_3; 0.65% K_2O; 0.3%
Na_2O. After casting and prior to conditioning, the specimens were cured in a fog-
room of 98% RH, at 20 °C for 28 days; after de-molding, electrical connections
were made to apply electrical current through cast-in electrodes (metal plates) on
the two opposite sides of each cube (Figs. 4.1 and 4.4). The electrodes also served
the purpose of measuring electrical resistivity (Figs. 4.1 and 4.4). For sealed conditions, the mortar specimens were sealed with bee wax to prevent water evaporation.
For water-submerged conditions, identical specimens were submerged in water-
containing vessels throughout the test.

4.3.2   Sample Designation and Current Regimes

The mortar specimens were cast in two main groups, differing in w/c ratio, i.e., 0.35
and 0.5. These two specimens' groups were presented by three subgroups: (1) control group – no DC current involved; (2) group "100 mA/m²" and (3) group "1 A/
m²", where DC current was relevant at the respective current levels.

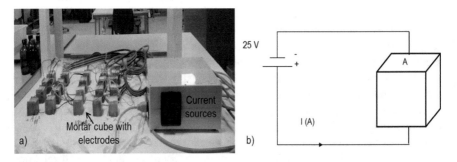

Fig. 4.4 (**a**) Experimental setup (sealed condition) for mortar cubes of w/c ratio 0.5 and 0.35; (**b**)
schematic representation for electrical current application through surface area A of the mortar
specimens

4.3.3 Methods

4.3.3.1 Mortar Electrical Resistivity

The electrical resistivity of the mortar cubes was measured using an alternating DC 2-pin method [27], where the "pins" are the metal electrodes (plates) with dimensions equal to the sides A of the mortar cubes, Fig. 4.4. To avoid polarisation effects during measurement for the water-submerged groups, the mortar cubes were taken out from the medium and cloth dried prior to testing. The resistance was measured by applying an alternating DC current of 1mA at a frequency of 1 kHz. An R-meter was used to record the electrical resistance of the mortar. For the "under current" regime (groups 100 mA/m^2 and 1 A/m^2), the resistance measurements were performed after current interruption of approximately 30 min. The electrical resistivity was calculated using Ohm's Law based on two probes measurements (as previously introduced in the Sect. 4.2.2).

4.3.3.2 Compressive Strength

Standard compressive strength tests were performed on the 40 mm × 40 mm × 40 mm mortar cubes at the hydration age of 28 days as initial measurement and later on after 3, 7, 14, 56, and 84 days of conditioning (i.e., 31, 42, 84, and 112 days of age). Three replicate mortar specimens were taken out from the conditioning setup and tested within a 5-min time interval.

4.4 Results and Discussion

4.4.1 Compressive Strength

Global mechanical properties of cement-based materials are of importance for the assessment of overall performance. In order to evaluate the influence of stray current flow on mechanical properties, compressive strength was recorded. Figure 4.5 shows the strength development of the 28-day cured mortar specimens of w/c ratio 0.5 and 0.35 in sealed and water submerged conditions. As expected, compressive strength increases with time and within cement hydration, which was relevant for all groups. The effect of w/c ratio can be clearly observed i.e. the specimens with w/c ratio of 0.35 presented higher compressive strength compared to those of w/c ratio 0.5, irrespective of exposure conditions and stray current effects. This trend of performance is well known and is generally due to a larger capillary porosity in the former case and a denser microstructure in the latter case [42].

   The effect of electrical current in sealed conditions is more obvious at later stages – after 56 days, where stabilisation of compressive strength for control cases was observed, whereas a still gradual increase was relevant for the "under current"

Fig. 4.5 Compressive strength of 28-day cured mortar specimens of w/c 0.35 and 0.5 as a function of hydration age in: **(a)** sealed conditions; **(b)** fully submerged in water conditions

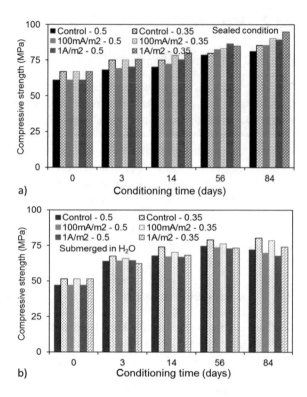

regimes. Increased compressive strength at the end of the test was especially relevant for the 1A/m^2 group at w/c ratio of 0.35 (Fig. 4.5a). This is a result of enhanced cement hydration at lower pore water content, leading to a more pronounced development of a rigid microstructure, if all these are compared to w/c ratio of 0.5 and control conditions.

In contrast to the sealed conditions, the compressive strength of water-submerged mortar decreased as a result of stray current (Fig. 4.5b). This was more pronounced towards the end of the test. Calcium ions leaching-out would result in depreciation of mechanical performance due to microstructural changes. This outcome is consistent with previous results on partly submerged in water specimens, where chemical analysis of the external medium, together with microstructural analysis supported the observed evolution of mechanical properties [27]. The effect of w/c ratio and stray current can be judged in parallel by comparing the difference in compressive strength between the specimens "under current" and the relevant control cases at identical age. Similarly to the sealed conditions, although with a reversed trend, the highest current density level resulted in the largest effect at the end of the test i.e. the lowest compressive strength for both w/c ratio 0.35 and 0.5 were recorded for mortar subjected to stray current of 1 A/m^2.

The trend of compressive strength development of the hereby tested mortar specimens at early and later stages is well in line with the development of electrical resistivity, as presented and discussed in what follows.

4.4.2   Electrical Resistivity

Electrical resistivity is a rapid and nondestructive testing method that provides an indication for the quality of concrete structures. The electrical resistivity of concrete can be also defined as the resistance of the matrix to water and aggressive ions penetration. It could also be an indication of properties development, when water and ion transport due to migration are concerned, e.g., when electrical current flows through a unit length or a unit cross-section of a concrete specimen.

The electrical current is "carried" by ionic charge, flowing through the concrete pore solution. Hence, water and ions migration in conditions of current flow will be involved in addition to diffusion-controlled transport (absorption or capillary suction are not considered). Enhanced ion and water flow would result in enhancement, or at least alteration, of cement hydration. Additionally, leaching-out for water-conditioned specimens, would contribute to microstructural changes and potentially be reflected in changes in electrical properties. All these would affect the development of electrical resistivity in conditions of current flow, if compared to control conditions. Next, the electrical resistivity values are affected by several factors including different concrete mixture (e.g. binder/cement type, water to cement ratio, aggregates, pozzolanic admixtures) and environmental conditions (e.g. temperature, humidity) [46].

For this experiment, w/c ratio and exposure conditions were variables, all other factors were maintained identical. The effect of electrical current was recorded with respect to w/c ratio and external medium and compared to identical control specimens in each test series.

For the sealed specimens, the electrical resistivity gradually increased with time (Fig. 4.6). A close to linear relationship was observed with the level of current density, the highest values were recorded for the $1 A/m^2$ regimes in both series of w/c ratios 0.35 and 0.5. As aforementioned, and as previously reported [27], this result is attributed to on-going cement hydration, minimized temperature loss and no interaction with external environment. Lower w/c ratio would additionally account for a denser microstructure and a reduction in the capillary pore volume of the mortar specimens. Therefore, the electrical resistivity values gradually increased with time due to the progress of cement hydration and subsequent development (densification) of the bulk microstructure.

Since the electrical current is carried by ionic flow through the pore solution of the mortar specimens, higher water/cement ratio results in an "easier" electrical current flow (i.e. low electrical resistance/resistivity). In contrast, the lower w/c ratio would impede electrical current flow (i.e. high electrical resistance/resistivity). If a correlation of factors is made e.g. w/c ratio and current flow effects on electrical properties of mortar, it is well seen that electrical resistivity increased for the specimens in conditions of electrical current, irrespective of the w/c ratio, Fig. 4.6c). In other words, for the specimen groups 100 mA/m^2 and 1 A/m^2, cement hydration was enhanced by elevated internal water and ion transport. This result in higher electrical resistivity over time – in the range of 480 to 530 Ohm.m, if compared to control conditions – approximately 350 Ohm.m towards the end of the test. Slightly higher

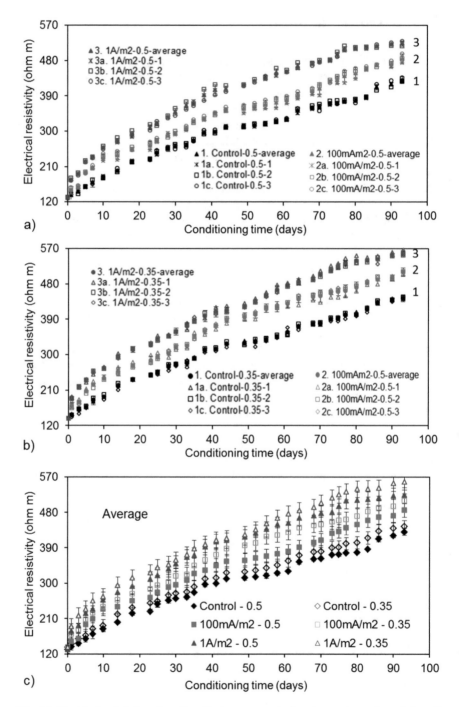

Fig. 4.6 Electrical resistivity values (3 replicates per group) of mortar specimens in sealed condition for w/c ratio 0.5 (**a**), w/c ratio 0.35 (**b**), and averaged data (**c**)

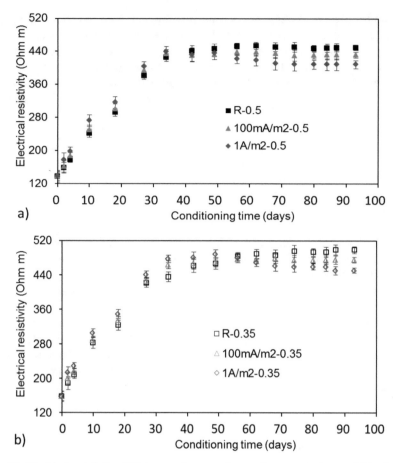

Fig. 4.7 Electrical resistivity of 28-day cured, water submerged mortar specimens of **(a)** w/c ratio 0.5 and **(b)** w/c ratio 0.35

values were recorded for the lower w/c ratio group (0.35) compared to the 0.5 w/c ratio group in both control and under current conditions (Fig. 4.6c), which was as expected and as previously discussed.

Figure 4.7 presents the evolution of electrical resistivity of the 28 days-cured mortar specimens, fully submerged in water. Similarly to sealed conditions, the effect of stray current was evaluated in parallel to the effect of varying w/c ratio and compared to control conditions of the same w/c ratio, i.e. w/c 0.5 (Fig. 4.7a) and w/c 0.35 (Fig. 4.7b). As can be observed, until ca. 40 days of conditioning (or ca. 70 days of age), the electrical resistivity for all specimens increased with time, irrespective of w/c ratio and conditions – Figs. 4.7a, b). This is logic and as expected, reflecting the maturity development of the cement matrix with time of hydration. After 40 to 55 days of conditioning, a subsequent increase and stabilization for the control specimens was recorded, but a decreasing trend for the "under current" conditions was observed. Increase or stabilization of electrical resistivity values follows the logic of continuous cement hydration with time and conditioning. The decrease of electrical resistivity for the water-conditioned specimens after longer treatment (> 50 days), was obvious, especially if compared to sealed specimens (Fig. 4.6). This result is linked to leaching-out and re-distribution of the pore network, which are not subject to discussion in this work. Some of these effects were previously reported for partly submerged in water specimens [27]. This negative effect of stray current can be well observed in the water-submerged group, reflected by the lowest electrical resistivity for specimens, subjected to the highest current density levels (Fig. 4.7). As can be also observed in both Fig. 4.7a) and Fig. 4.7b), the negative effect of stray current was not determined by w/c ratio i.e. was relevant at comparable levels in both cases of w/c 0.35 and w/c 0.5, despite the lower amount of pore water at the w/c ratio of 0.35, compared to that of w/c ratio 0.5.

The results from electrical resistivity measurements are well in line with the recorded compressive strength in both sealed and water-submerged conditions (Fig. 4.5). Increasing electrical resistivity and higher compressive strength for sealed conditions were the result from stray current, where no concentration gradient with external medium was relevant. In contrast, decrease in electrical and mechanical properties was observed for water-treated specimens, more pronounced for the highest current density level of 1 A/m^2.

Following the above discussed experimental results and the outlined in Section 4.2. methodology, the ageing factor for sealed mortar in the hereby tested conditions was determined. Next, the apparent diffusion coefficients were calculated and results discussed in view of the effect of stray current on mortar specimens of different w/c ratio.

4.4.3   Ageing Factors Determination

The recorded electrical resistivity values, as presented in Figs. 4.6 and 4.7, were employed in Eq. (4.5), which through curve fitting was used to derive the exponents q. The as derived exponent q values were later on used to calculate the ageing factors n, following Eq. (4.6). Table 4.1 summarizes the obtained values for each group of specimens. An example of the fitting results for the sealed specimens are presented in Fig. 4.8 for w/c ratio 0.35 (Fig. 4.8a) and w/c ratio 0.5 (Fig. 4.8b). Curve fitting of Eq. (4.5) for water-submerged conditions was not possible, since on one hand the obtained error was very large and not presenting meaningful results. On the other hand, it is logic that the same function cannot be employed for two very

Table 4.1 Summarized q values and calculated aging factor n (Eq. 4.6) for mortar, cast from CEMI 42.5 N and cured for 28 days

Parameters	Sealed experiments – w/c 0.5			Sealed experiments – w/c 0.35		
	Control	100 mA/m^2	1 A/m^2	Control	100 mA/m^2	1 A/m^2
Values of q	0.295	0.310	0.328	0.306	0.321	0.337
Aging factor (n)	0.369	0.388	0.410	0.383	0.401	0.421

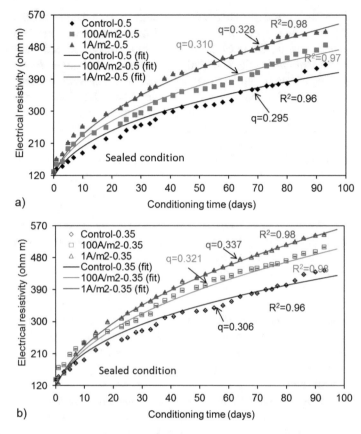

Fig. 4.8 Electrical resistivity values of mortar specimens in sealed condition and aging factor "*q*" obtained from fitting curve using Eq. (4.5) with w/c ratio 0.5 (**a**) and w/c ratio 0.35 (**b**)

different exposure conditions. Therefore, the application of the reported in literature analytical approaches for ageing factors determination, based on electrical resistivity, were only applied to the sealed specimens in this work.

As seen in the Fig. 4.8, the exponent *q* values increased with increasing the level of electrical current for both w/c ratio 0.5 and 0.35. The exponent *q* values for control conditions was in the range of 0.369 - 0.383, while for the 100 mA/m^2 and 1A/m^2 groups a range of 0.388 - 0.421 was recorded. It was also recorded that the exponent *q* values for the mortar groups of w/c ratio 0.35 was slightly higher than that, derived for w/c ratio 0.5. Although the difference was not significant, the result can be attributed to a potentially more pronounced effect on cement hydration, and a more substantial pore refinement for the mortar specimens of w/c ratio 0.35 (lower amount of capillary water initially present if compared to w/c ratio 0.5) resulting in increase of electrical resistivity. Additionally, lower w/c ratio tends to result in higher electrical resistivity due to a lower content of the liquid phase.

Table 4.1 summaries exponent q values as derived from the electrical resistivity records and the ageing factor n as calculated from the relationship between the exponent q and n, using Eq. (4.6). From Eq. (4.6) it can be deduced that the exponent q, derived from resistivity is slightly lower than n, n obtained from diffusion coefficient. This is due to the fact that both electrical resistivity (ρ) and apparent diffusion coefficient (D_{ap}) evolve differently with time [17, 36]. Among other phenomena, the values of electrical resistivity (ρ), represent the evolution of microstructural properties (i.e. pore network and its connectivity) due to cement hydration and/or reactions with pozzolanic material, mineral admixtures, etc. Electrical resistivity changes can reflect variation in diffusion coefficients as well, since diffusivity is determined by microstructure development, including chloride binding capacity in the bulk matrix. The diffusion coefficient Dap, however, also reflects the surface chloride binding and chloride surface concentration with time. Although the difference between q and n values (Table 4.1) was not well pronounced, the variation of derived values can be attributed to the influence of stray current on ions and water migration. These in turn affect microstructural development and the evolution of mechanical and electrical properties, respectively.

As seen in the Table 4.1, both n and q values changed with water-to-cement ratio and the level of stray current flow. As expected, the aging factor n increased with decreasing of water-to-cement ratio and with increasing electrical resistivity values. As aforementioned, lower water-to-cement ratio results in higher electrical resistivity, as stated by MacDonald and Northwood [47], and the variation of electrical resistivity is caused by changing in porosity, determined by water content and the water-to-cement ratio.

For the experimental conditions of this work, the aging factor was obviously not only depending on the w/c ratio, but was also affected by the stray current flow. The simultaneous effect of stray current flow and water-to-cement ratio can be deduced from the results in Table 4.1. It can be seen that the aging factor increased with increasing the level of stray current flow for both w/c ratio 0.5 and 0.35. The aging factor for specimens of w/c ratio 0.35 tends to have higher values than that for specimens of w/c ratio 0.5. At lower w/c ratio, the stray current flow had a larger effect in view of increasing the temperature of a lower amount of pore water. As a consequence, the rate of cement hydration increased faster for specimens of w/c ratio 0.35, compared to these of w/c ratio 0.5. This in turn led to a more rapid pore refinement, or densification of cement bulk matrix, in specimens of w/c 0.35. However, if control and under current conditions are compared within the groups of same w/c ratio (Table 4.1), the following can be observed: compared to control conditions, the aging factor increased with 5.1% and 11.1% at the level of current densities 100 mA/m^2 and 1 A/m^2, respectively, for the groups of w/c ratio 0.5. In contrast, for the groups of w/c ratio 0.35, the aging factor increased with about 4.7% and 10% at the current density levels of 100 mA/m^2 and 1 A/m^2. Although the overall aging factor for w/c ratio 0.35 was slightly higher than that for w/c ratio 0.5, the stray current flow contributed to the aging factor changes to a higher extent at w/c ratio 0.5, if under current regimes are compared to control conditions in one and the same w/c ratio group.

Table 4.2 Aging factor n for OPC mixtures from literatures

w/c	n-values	References
0.5	0.320	Stanish [48]
0.5	0.178	Zhuqing Yu [24]
0.45	0.264	Bamforth [23]
0.45	0.275, q = 0.22	Andrade [11]
0.45	0.236	Dalen [31]
0.45	0.182	Maes [49]
0.4	0.350	F. Presuel-Moreno [13]

The derived exponent q and n values in this study are well in line with values reported in the literature, especially if conditions similar to the hereby tested control series are considered [11, 22, 23]. As can be observed from Table 4.2 (e.g. ref. [11] and [48] refer to sealed conditions, ref. [13] – to immersed, ref. [23] – to splash zone, ref. [24] – to lab), the ageing factor (n) determined by both electrical resistivity-based and diffusion coefficient-based approach as previously reported, are in the range of the hereby derived ones (Table 4.1). This outcome denotes for a valid approach towards retrieving reliable results for the present investigation, on one hand. On the other hand, the effect of stray current on ageing factors and maturity level of cement-based materials is apparently possible to be quali- and quantified following the same approach.

4.4.4 Diffusion Coefficient

By considering the previously introduced theoretical background and employing the aging factor n as presented in the Table 4.1, the apparent diffusion coefficient as function of time can be predicted, using Eq. (4.3). Graphically this is presented in Figs. 4.9 and 4.10. Figure 4.9 depicts a comparison of the apparent diffusion coefficient (D_{ap}) as derived from the obtained aging factors n in this study (curve "control 0.5") and D_{ap} as reported in literature for control conditions. As seen in the Fig. 4.9, there is a good agreement between the D_{ap} values, predicted based on calculation from the aging factor n and these derived from direct measurements of apparent diffusion coefficient. This outcome can be further used to predict the influence of stray current flow on apparent diffusion coefficient of cement-based materials. Figure 4.10 depicts the evolution of D_{ap} for the under current subgroups 100 mA/m^2 and 1 A/m^2 in comparison to the control group in this study. As expected, the apparent diffusion coefficient did not change significantly at early stages, which holds for both groups of w/c ratio of 0.5 and 0.35. A gradual decrease, however, was observed at later stages with decreasing the w/c ratio (i.e., D_{ap} of w/c ratio 0.35 is lower than that of w/c ratio 0.5).

As far as the effect of stray current is concerned, in both groups of w/c ratios of 0.35 and 0.50, a decrease of D_{ap} was observed over time as an indirect effect of the stray current. This was significantly more pronounced for the higher current density group (1 A/m^2) in combination with lower w/c ratio (0.35).

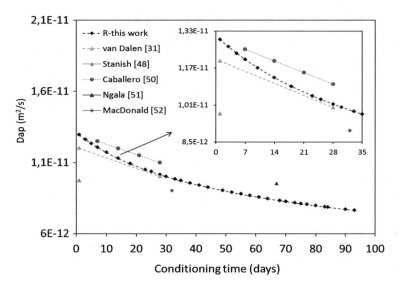

Fig. 4.9 Prediction of apparent diffusion coefficient calculated using Eq. (4.3) based on aging factor (**n**), as derived from electrical resistivity records, compared to reported literature results

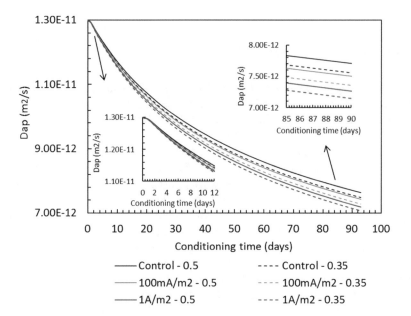

Fig. 4.10 Apparent diffusion coefficient of mortar specimens subject to stray current flow calculated using Eq. (4.3) based on aging factor (**n**) derived from electrical resistivity records

Obviously, the different levels of stray current density and w/c ratio – as separately evaluated and as a synergetic effect – exert influence on the maturity levels and aging factors of cement-based materials. Electrical resistivity values can be used as an alternative of diffusion coefficients within the determination of the aging factor of cement-based materials. Moreover, the obtained results, using this alternative approach, are in a very good agreement with results, where values from chloride diffusion tests were used.

In a "reversed" approach, the aging factors, as derived based on electrical resistivity, can be used to determine diffusion coefficients and predict the development of apparent chloride diffusivity. This approach can be used further to predict service life of cement-based materials due to stray current flow. This study discussed only relatively lower levels of current flow, which result in positive effects, e.g., densification of the bulk matrix, increased compressive strength. Should be noted that in practical situations, this is not necessarily the case and the level of stray current can be significant and leading to negative side effects. For example, if combined with environmental conditions as level of humidity and chloride content, stray current can cause leaching out of calcium-baring phases and reduced global properties and performance would be relevant. This is especially the case when enhanced water and ion migration are involved, as within electrical current flow through a cement-based matrix.

Therefore, further research is necessary in order to define a threshold value for positive and/or negative effects of electrical current flow (stray current respectively) on cement-based material properties and performance within service life. In the following section, the concept of applying the above considerations to (potentially practical) cases of significantly larger current density levels is outlined. Also presented is an example for the negative effect of stray current in relevant environmental conditions and in view of aging factors determination.

4.5   Conclusions

This chapter discussed the influence of stray current flow on material properties and performance of 28-day cured cement-based materials in two distinct environmental conditions. The ageing factor, as a parameter linked to durability of cement-based materials, was determined from electrical resistivity development for sealed conditions. Based on the experimental results and analytical approach, the following conclusions can be drawn:

The ageing factor for control specimens in sealed conditions, obtained from electrical resistivity development over time, was in the range of those published by other researchers. This indicates that this method is potentially reliable and can be applied also to determine the influence of stray current flow on cement-based materials. The results show that the influence of stray current not only depends on the level of current flow, but is also largely determined by the relevant environmental conditions.

In sealed conditions, the ageing factor (n) obtained from electrical resistivity increased with the level of stray current flow and with lowering the w/c ratio, as also indicated by increase in compressive strength development. The ageing factor for water-submerged specimens was not possible to be determined using the same approach. However, the real experimental data shows the negative effect of stray current in view of decreased electrical properties and mechanical performance.

The approach to determine the ageing factor from electrical resistivity development may be useful to determine the apparent diffusion coefficient of cement based materials to a certain extent and in controlled environment. Deriving ageing factors for various exposure conditions, however, needs to consider other factors and materials development over time. The hereby discussed analytical functions were found to be not directly applicable for water-submerged specimens in view of the effects of stray current.

Acknowledgements The financial support from Directorate General of Higher Education Ministry of Education Republic of Indonesia is gratefully acknowledged. The authors would like to thank technicians of Microlab, Section of Material and Environment, Delft University of Technology for supporting an experimental setup.

References

1. Audenaert K, Yuan Q, de Schutter G (2010) On the time dependency of the chloride migration coefficient in concrete. Constr Build Mater 24(3):396–402
2. Tang L (1999) Concentration dependence of diffusion and migration of chloride ions: Part 2. Experimental evaluations. Cem Concr Res 29(9):1469–1474
3. Tang L, Gulikers J (2007) On the mathematics of time-dependent apparent chloride diffusion coefficient in concrete. Cem Concr Res 37(4):589–595
4. Tang L, Nilsson LO (2004) On relationships between different chloride diffusion and/or migration coefficients in concrete. In: Proceedings of the 3rd RILEM workshop on testing and modelling the chloride ingress into concrete. RILEM Publications Sarl, Bagneux, pp 317–328
5. Yuan Q, De Schutter G, Shi C, Audenaert K (2008) The relationship between chloride diffusion and migration coefficients in concrete. In: 1st international conference on microstructure related durability of cementitious composites, 13–15 Oct 2008, Nanjing
6. Mingzhong Z (2013) Multiscale lattice Boltzman-finite element modelling of transport properties in cement-based materials. PhD thesis, Delft University of Technology
7. NORDTEST NT BUILD 443 (1995) Accelerated chloride penetration, Finland
8. NORDTEST NT BUILD 492 (1999) Chloride migration coefficient from non-steady state migration experiments, Finland
9. NORDTEST NT BUILD 355 (1997) Chloride diffusion coefficient from migration cell experiments, Finland
10. Mangat P, Molloy B (1994) Prediction of long term chloride concentration in concrete. Mater Struct 27(6):338–346
11. Andrade C, Castellote M, d'Andrea R (2011) Chloride aging factor of concrete measured by means of resistivity. Proceedings of the XII-international conference on durability of building materials and components, Porto
12. Andrade C, Castellote M, d'Andrea R (2011) Measurement of ageing effect on chloride diffusion coefficients in cementitious matrices. J Nucl Mater 412:209–216
13. Presuel-Moreno F, Wu Y-Y, Liu Y (2013) Effect of curing regime on concrete resistivity and aging factor over time. Constr Build Mater 48:874–882

14. Millard SG, Gowers KR (1992) Resistivity assessment of in-situ concrete: the influence of conductive and resistive surface layers. Proc Inst Civil Eng Struct Build 94(4):389–396
15. Brooks AG, Schindler AK, Barnes RW (2007) Maturity method evaluated for various cementitious materials. J Mater Civ Eng 19:1017–1025
16. Bentz DP, Garboczi EJ (1999) Effects of cement particle size distribution on performance properties of Portland cement-based materials. Cem Concr Res 29(10):1663–1671
17. Brown PW, Dex S, Skalny JP (1993) Porosity/permeability relationships. Parts of 'Concrete microstructure porosity and permeability'. Roy DM, Brown PW, Shi D, Scheetz BE, May W. Strategic Highway Research Program, National Research Council, Washington, DC 1993. SHRP-C-628
18. Tang L, Nilsson LO (1992) Chloride diffusivity in high strength concrete at different ages. Nor Concr Res 11(1):162–171
19. Maage M, Helland S, Poulsen E, Vennesland O, Carlsen JE (1996) Service life prediction of existing concrete structures exposed to marine environment. ACI Mater J 93(6):602–608
20. Gulikers J (2006) Pitfalls and practical limitation in probabilistic service life modeling of reinforced concrete structures. In: Proceedings of the Eurocorr 2006, Maastricht 25–28 Sept 2006
21. Castellote M, Andrade C, d'Andréa R (2009) The use of resistivity for measuring aging of chloride diffusion coefficient. In: Proceedings RILEM TC 211-PAE international conference – concrete in aggressive aqueous environments, Toulose
22. Gehlen C (2000) Probabilistische Lebensdauerbemessung von Stahlbetonbauwerken, Dafst 510. Beuth Verlag, Berlin
23. Bamforth PB (1999) The derivation of input data for modelling chloride ingress from eight year UK coastal exposure trials. Mag Concr Res 51(2):87–96
24. Yu Z (2015) Microstructure development and transport properties of Portland cement-fly ash binary systems – in view of service life predictions. PhD thesis, Delft University of Technology, Delft
25. DuraCrete (2000) Probabilistic performance based durability design of concrete structures. Contract BRPR-CT95–0132, Project BE95–1347. Document BE95–1347/R17
26. CUR-Bouw&Infra (2009) Duurzaamheid van constructief beton met betrekking tot chloride-geïnitieerde wapeningscorrosie – Leidraad voor het formuleren van prestatie-eisen-Achtergrondrapport (in Dutch), Tu Delft, 19 Mei 2009
27. Susanto A, Koleva DA, Copuroglu O, van Beek C, van Breugel K (2013) Mechanical, electrical and microstructural properties of cement-based materials in conditions of stray current flow. J Adv Concr Technol 11(3):119–134
28. Andrade C., d'Andrea R. Electrical resistivity as microstructural parameters for the modelling of service life of reinforced concrete structures, 2nd international symposium on Service Life Design for Infrastructure, October 2010, Delft, RILEM proceedings
29. Singh Y (2013) Electrical resistivity measurements: a review. Int J Mod Phys Conf Series 22:745–756
30. Wenner F (1915) A method for measuring earth resistivity. Bull Bureau Stand 12:469–478
31. van Dalen Sander M (2005) Experimenteel onderzoek naar de RCM-methode (In Dutch). Master thesis, Delft University of Technology
32. Naish CC, Harker A, Carney RFA (1990) Concrete inspection: interpretation of potential and resistivity measurements. In: Page CL, Treadaway KWJ, Bamforth PF (eds) Corrosion of reinforcement in concrete. SCI, London, pp 314–332
33. Millard SG (1991) Reinforced concrete resistivity measurement techniques. Proc Inst Civil Eng 2:71–88
34. Telford WM, Telford WM, Geldart LP, Sheriff RE (1990) Applied geophysics. Cambridge University Press, New York
35. Polder R, Andrade C, Elsener B, Vennesland O, Gulikers J, Weidert R, Raupach M (2000) Rilem TC 154-EMC: electrochemical techniques for measuring metallic corrosion – test methods for on site measurement of resistivity of concrete. Mater Struct Mater Constr 33:603–611
36. Feliu S, Andrade C, Gonzalez JA, Alonso C (1996) A new method for in situ measurement of electrical resistivity of reinforced concrete. Mater Struct 29:362–365

37. Susanto A, Koleva DA, van Breugel K (2015) The effect of temperature rise on microstructural properties of cement-based materials: correlation of experimental data and a simulation approach. In: Mori G, Hribernik B, Stellnberger KH, Dworak Y (eds) Proceeding of the European corrosion congress, EUROCORR 2015 (pp 1–10). s.l.: European Federation of Corrosion

38. Susanto A, Koleva DA, van Breugel K, Koenders EAB (2014) Modelling the effect of electrical current flow on the hydration process of cement-based materials. In: Mangabhai RJ, Bai Y, Goodier CI (eds) Extended abstract: young researchers' forum II: construction materials. University College London, London, pp 41–48

39. Susanto A, Koleva DA, van Breugel K (2014) Altered cement hydration and subsequently modified porosity, permeability and compressive strength of mortar specimens due to the influence of electrical current. In: Lazzari L, Fedrizzi L, Mol A (eds) Proceedings of the European corrosion congress (EUROCORR 2014). European Federation of Corrosion, Oxford, pp 1–10

40. Heritage I (2001) Direct electric curing of mortar and concrete. Napier University, Edinburgh

41. Bredenkamp S, Kruger D, Bredenkamp GL (1993) Direct electric curing of concrete. Mag Concr Res 45(162):71–74

42. Brooks AG, Schindler AK, Barnes RW (2007) Maturity method evaluated for various cementitious materials. J Mater Civ Eng 19:1017–1025

43. Abdel-Jawad YA (2006) The maturity method: Modifications to improve estimation of concrete strength at later ages. Constr Build Mater 20:893–900

44. Wilson JG, Gupta NK (2004) Equipment for the investigation of the accelerated curing of concrete using direct electrical conduction. Measurement 35:243–250

45. Susanto A, Koleva DA, van Breugel K (2014) DC current-induced curing and ageing phenomena in cement-based materials. In: van Breugel K, Koenders EAB (eds) Proceedings of the 1st international conference on ageing of materials and structures, AMS'14 (pp 562–568), 26–28 May 2014, Delft

46. Sengul O (2013) Factors affecting the electrical resistivity of concrete, Nondestructive Testing of Materials and Structures, RILEM Bookseries 6, doi 10.1007/978–94–007-0723-8-38, © RILEM 2013

47. MacDonald, K. A., and Northwood, D. O., Rapid estimation of water-cementitious ratio and chloride ion diffusivity in hardened and plastic concrete by resistivity measurement, S-P-191, M. S. Khan, American Concrete Institute, Farmington Hills, 1999, pp 57–68.

48. Stanish K, Thomas M (2003) The use of bulk diffusion tests to establish time-dependent concrete chloride diffusion coefficients. Cem Concr Res 33:55–62

49. Maes M, Caspeele R, Van den Heede P, De Belie N (2013) Influence of sulphates on chloride diffusion and the effect of this on service life prediction of concrete in a submerged marine environment. In: Strauss A, et al Life-cycle and sustainability of civil infrastructure systems proceedings of the third international symposium on life-cycle civil engineering (IALCCE'12), Vienna, 3–6 Oct, 2012. pp 899–906

50. Caballero J, Polder RB, Leegwater GA, Fraaij ALA (2012) Chloride penetration into cementitious mortar at early age. Heron 57(3):185–196

51. Ngala VT, Page CL, Parrott LJ, Yu SW (1995) Diffusion in cementitious materials: II further investigations of chloride and oxygen diffusion in well-cured OPC and OPC/30%PFA pastes. Cem Concr Res 25(4):819–826

52. MacDonald KA, Northwood DO (1995) Experimental measurements of chloride ion diffusion rates using a two-compartment diffusion cell: effects of material and test variables. Cem Concr Res 25(7):1407–1416

Chapter 5
Electrochemical Tests in Reinforced Mortar Undergoing Stray Current-Induced Corrosion

Zhipei Chen, Dessi A. Koleva, and Klaas van Breugel

Abstract Stray current, arising from direct current electrified traction systems, further circulating in nearby reinforced concrete structures may initiate corrosion or accelerate existing corrosion processes on the steel reinforcement. In some extreme conditions, corrosion of the embedded steel will occur at very early stage. One of the significant consequences is loss of bond strength and premature failure of the steel-matrix interface. This plays an important role for the integrity of a structure during its designed service life.

In this work, the level of stray current was set at 0.3 mA/cm^2, applied as an external DC electrical field. This level of stray current was chosen based on preliminary calculations on expected corrosion damage, i.e., in view of material loss at the level of 10% weight loss of the steel rebar (analytically calculated via Faraday's law). The investigated reinforced mortar specimens were cured for 24 h only and then conditioned in chloride-free and chloride-containing environment. The evolution of steel electrochemical response in rest (no stray current) and under current conditions was monitored for approx. 240 days via OCP (Open Circuit Potential), LPR (Linear Polarization Resistance), EIS (Electrochemical Impedance Spectroscopy) and PDP (Potentio-dynamic Polarization).

The results show that the effect of stray current on concrete bulk matrix properties, together with steel corrosion response, is significantly determined by the external environment, as well as by the level of maturity of the cement-based bulk matrix.

For chloride-free environment the effect of the chosen stray current level was not significant, although lower corrosion resistance of the steel rebars was

Z. Chen (✉) • K. van Breugel
Faculty of Civil Engineering and Geosciences, Department Materials and Environment,
Delft University of Technology, Stevinweg 1, 2628, CN, Delft, The Netherlands
e-mail: z.chen-1@tudelft.nl

D.A. Koleva
Delft University of Technology, Faculty of Civil Engineering and Geosciences,
Section Materials and Environment, Stevinweg 1, 2628, CN, Delft, The Netherlands

© Springer International Publishing AG 2017　　　　　　　　　　　　　　　　83
L.E. Rendon Diaz Miron, D.A. Koleva (eds.), *Concrete Durability*,
DOI 10.1007/978-3-319-55463-1_5

recorded after longer exposure of ~240 days, compared to control conditions. In fact, even positive effects of the stray current were observed in the early stages, i.e., until 28 days of age: stray current flow through a fresh (non-mature) cement matrix led to enhanced water and ion transport due to migration. The result was enhanced cement hydration, consequently environment, assisting a more rapid stabilization of pore solution and steel/cement paste interface. In chloride-containing external medium, steel corrosion was a synergetic effect of both de-passivation due to chloride ions in the medium and stray current effects. Corrosion acceleration solely due to the stray current flow in chloride-containing medium cannot be claimed for the chosen current density levels and the duration and conditions of the experiment.

What can be concluded is that the effect of stray current for both chloride-free and chloride-containing conditions is predominantly positive in the initial stages of this test. The expected negative influence towards corrosion acceleration was observed after a prolonged treatment, when a stable maturity level of the cement-based matrix was at hand. This also means that the properties of the cementitious material in reinforced cement-based system are of significant importance and largely determine the electrochemical state of the steel reinforcement.

Keywords Stray current • Steel corrosion • Mortar • Electrochemical tests

5.1 Introduction

In electrical current powered traction systems, such as electrical trains, tram systems or underground trains, the current drawn by the vehicles returns to the traction power substation through the running rails. This concept, besides forming part of the signaling circuit for controlling the train movements, is used as the current return circuit path, together with return conductors [1–5]. However, owing to the longitudinal resistance of the rails and their imperfect insulation to the ground, part of the return current leaks out from the running rails, i.e., stray current forms. The leaking-out stray current returns to the traction power substation through the ground and underground metallic structures (such as steel rebars in concrete, pipelines), as schematically shown in Fig. 5.1 [6, 7]. Stray direct currents (DC) are known to be more dangerous than stray alternating currents (AC) [3, 8, 9]. Such situation can be potentially very harmful for underground reinforced concrete structures for example, where despite the high concrete resistivity the steel reinforcement provides a good conductive path. The circulation or flow of stray current through reinforced concrete structures may initiate corrosion or significantly accelerate already existing corrosion processes on the steel reinforcement [1].

For a reinforced concrete structure, buried near a rail track, a DC stray current leakage will initiate or enhance steel corrosion as follows: at the point where the

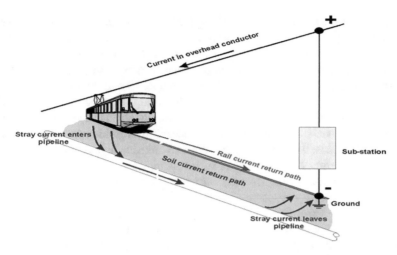

Fig. 5.1 Example of stray current from a railway line picked up by pipeline [10]

stray current "enters" the reinforcement, a cathodic reaction occurs (predominantly oxygen reduction for this environment):

$$\frac{1}{2}O_2 + H_2O + 2e^- \rightarrow 2OH^- \tag{5.1}$$

The point where the stray current flows out from the reinforcement into the external environment (e.g., concrete, soil), anodic reaction occurs which is the process of metal dissolution:

$$Fe \rightarrow Fe^{2+} + 2e^- \tag{5.2}$$

In some extreme conditions (for instance, newly built reinforced concrete structures nearby seaside and/or railways), stray current-induced corrosion of the steel reinforcement could occur at early age and can be significant. This is because of the lower level of maturity of the cement-based matrix in young concrete, which would in turn facilitate the stray current-induced corrosion on the steel surface. Lower electrical resistivity, higher porosity, permeability, diffusivity, etc. of the concrete matrix would lead to enhanced aggressive ions penetration in addition to elevated ion and water migration due to the stray current flow. On the one hand, for a non-matured cement-based matrix and at early stages of cement hydration, elevated ion and water migration can have a positive effect in view of enhanced cement hydration and chloride-binding capacity of the bulk material [11, 12]. However, at later stages and for an already developed bulk matrix microstructure, stray current flow can affect the pore space and hydration products' network due to leaching-out of calcium baring phases, coarsening of the pore structure, etc. [13]. These will in turn enhance the purely electrochemical reactions, related to the corrosion process on the steel reinforcement. One of the significant consequences is the premature failure of the steel-matrix interface, i.e., significantly reduced bond strength, which plays important role for the integrity of a structure during the subsequent service life. These phenomena lead frequently to early deterioration and eventually to situations of high risk in view of durability and serviceability of reinforced concrete structures. By all means, the above concerns are also linked to high cost for maintenance and repair, targeting safe operation of such structures.

This work aimed to simulate stray current-induced corrosion for reinforced concrete and investigate the electrochemical response of the steel reinforcement in comparison to control (stray current-free) conditions. In order to account for the significant effect of environmental conditions, mainly the presence of chloride ions as a corrosion accelerator, the stray current conditions were simulated in both chloride-free and chloride-containing medium. The level of stray current was set at 0.3 mA/cm² (through the application of external DC electrical field). This level of current density was chosen in a way to account for a hypothetic 10% weight loss of the steel rebar.

The overall aim of the research project on stray-current corrosion, this work being an initial part of it, is to correlate the electrochemical response of steel (corrosion state) with bond strength at the steel/cement paste interface and bulk matrix properties. This chapter focuses only on the initial results from electrochemical monitoring of reinforced mortar specimens in the time frame of the test of 243 days. Several types of general and more sophisticated electrochemical tests are presented and discussed in view of the effect of stray current on the corrosion state of steel reinforcement. Additionally, the development and alterations in the properties of the bulk matrix were indirectly evaluated, owing to the possibility of one of the electrochemical non-destructive techniques (i.e., electrochemical impedance spectroscopy) to provide qualitative and quantitative information for both steel- and cement-based material properties.

5.2   Experimental

5.2.1   Materials and Specimen Preparation

Reinforced mortar prisms of 40 mm × 40 mm × 160 mm were cast from Ordinary Portland cement CEM I 42.5 N and normed sand. The water-to-cement (w/c) ratio was 0.5; the cement-to-sand ratio was 1:3. Construction steel (rebar) FeB500HKN (d = 6 mm) with exposed length of 40 mm was centrally embedded in the mortar prisms (Fig. 5.2). Prior to casting, the steel rebars were cleaned electrochemically (cathodic treatment with 100 A/m² current density, using stainless steel as anode) in

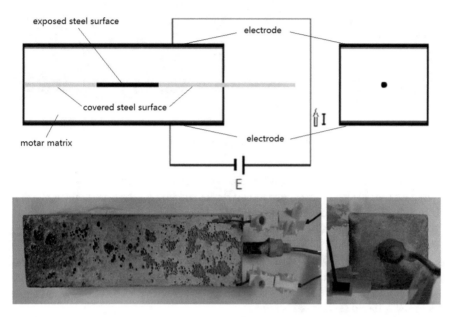

Fig. 5.2 Experimental setup and position of electrodes for stray current supply

a solution of 75 g NaOH, 25 g Na_2SO_4, 75 g Na_2CO_3, reagent water to 1000 mL, according to ASTM G-1 [14, 15]. The schematic presentation of the specimens' geometry is depicted in Fig. 5.2: the two ends of the rebars were covered by heat-shrinkable tube, in order to restrict the exposed surface area.

The experimental setup and electrodes' configuration for supplying stray current (with the level of 0.3 mA/cm^2) are as presented in Fig. 5.2: two Ti electrodes (MMO Ti mesh, 40 mm × 160 mm) cast-in within samples preparation served as terminals for stray current application. When the stray current supply was interrupted (min 24 h before electrochemical tests), the Ti electrodes served as counter electrode in a general 3-electrode setup, where the rebar was the working electrode and an external SCE (Saturated Calomel Electrode) served as a reference electrode.

5.2.2 Curing Conditions

After casting and prior to demoulding and conditioning, all specimens were cured in a fog room (98% RH, 20 °C) for 24 h. The specimens were lab conditioned after demoulding. All specimens were immersed with two-third of height in water or 5% NaCl solution. Table 5.1 summarizes the relevant conditions and specimens' designation.

Table 5.1 Details of
conditioning regimes

Group	External environment	Current supply
R	Water (2/3)	–
C	5% NaCl (2/3)	–
S	Water (2/3)	Stray current
CS	5% NaCl (2/3)	Stray current

5.2.3 Level of Stray Current Regime: Considerations

Besides present work, a series of side-by-side tests have also been performed to compare the effects of stray current and direct anodic polarization on the corrosion behavior of steel embedded in mortar specimens, due to the fact that some current state-of-the-art generally reports on simulating stray current corrosion through anodic polarization.

For a hypothetic 10% weight loss of the steel rebar reduced by anodic polarization, the applied current density (i.e., level of anodic polarization) can be expressed as a function of duration of the corrosion process and predefined corrosion degree as follows, based on Faraday's law:

$$i = \frac{ZFr\rho\eta_s}{2At} \tag{5.3}$$

where t is time of corrosion (for the feasibility of calculation, the duration of 28 days has been chosen preliminarily, 28 days =2,419,200 s), Z is the valence of the metal ions taking part in the anodic reaction, which is 2 in this case (iron), F is Faraday's constant ($F = 96,500$ As), r is the radius of corroded bar (0.3 cm), ρ is the density of iron ($\rho= 7.87$ g/cm^3), η_s is the mass loss ratio (10%), A is the atomic mass of iron ($A = 56$ g) and i is the impressed current density (A/cm^2).

Based on the calculation, the level of anodic polarization corresponding to 10% mass loss is 0.1744 mA/cm^2. Considering the fact that the supplied anodic current may not all take part in the oxidation of iron in mortar matrix, part of it may be involved in other reactions. So for this consideration, the final chosen current level was increased and set at 0.3 mA/cm^2.

For the convenience of comparison between stray current and anodic polarization, levels of stray current and anodic polarization are both set at 0.3 mA/cm^2.

5.2.4 Testing Methods

The electrochemical measurements were performed at open circuit potential (OCP) for all specimens, using SCE as reference electrode (as above specified, the counter electrode was the initially embedded MMO Ti). The used equipment was Metrohm Autolab (Potentiostat PGSTAT302N), combined with a FRA2 module.

Electrochemical impedance spectroscopy (EIS) was employed in the frequency range of 50 kHz − 10 mHz, by superimposing an AC voltage of 10 mV (rms). Linear polarization resistance (LPR) was performed in the range of ±20 mV vs OCP, at a scan rate of 0.1 mV/s. Both EIS and LPR tests were performed after 3, 7, 14, 28, 56, 141, 215 and 243 days of conditioning. Prior to each EIS and LPR test, a 24 h depolarization (potential decay) was recorded to assure stable OCP after stray current interruption. During the depolarization process and within electrochemical tests, the specimens were immersed fully in the relevant medium. The control samples (Group R) were also monitored according to the above test procedure. At the end of conditioning (at the age of 243 days), PDP (potentio-dynamic polarization) was performed in the range of −0.15 V to +0.90 V vs OCP at a scan rate of 0.5 mV/s, in order to additionally collect information for the electrochemical state of the steel reinforcement.

5.3 Results and Discussion

The outcomes from electrochemical tests are presented and discussed starting from non-destructive tests – e.g., OCP records and EIS measurements throughout the experiment, followed by results from polarizations tests (PDP) at the end of the test. A relatively larger portion of the below discussion is on OCP records in view of considerations for the expected and unexpected electrochemical response of the embedded steel in this work. EIS results are mainly discussed as a qualitative assessment in view of bulk matrix properties and steel surface active/passive state, however, also including a global quantitative assessment of corrosion state over time. Finally, the PDP tests support the results and discussion in the preceding OCP and EIS sections.

5.3.1 Open Circuit Potentials and Polarization Resistance

5.3.1.1 General Considerations OCP Readings and R_p Values

The general meaning of OCP values and their evolution in time accounts for the global corrosion state of the steel reinforcement – active or passive. Figure 5.3a also indicates the threshold for passive to active state for steel in cement-based materials, as defined by accepted standards and criteria [10]. The OCP values are used for a qualitative assessment only of corrosion state. In contrast, R_p values are used for a quantitative assessment, through calculating corrosion current by employing the Stern-Geary equation, i.e., $i_{corr} = B/R_p$ [16]. The R_p value is experimentally derived, whereas the constant B can be either experimentally derived or reported values for passive (B = 52 mV/dec) or active (B = 26 mV/dec) state can be employed [16]. Since R_p is inversely proportional to the corrosion current, quantification of

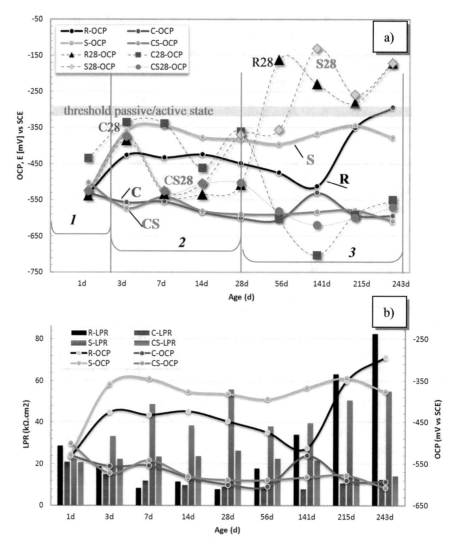

Fig. 5.3 (**a**) OCP evolution with time (1 –243 days) for specimen groups: *R* Reference, *C* corrod-ing (NaCl medium), *S* Stray Current, *CS* Corroding (NaCl) + Stray Current – all after *24 h curing* and subsequently conditioned until 243 days; (**b**) evolution of R_p values and correlation to OCP values over time for specimen groups *R, C, S* and *CS*. For comparative purposes, Fig. 5.3a also depicts the OCP evolution for *R, C, S* and *CS* specimens from parallel groups, cured *for 28 days* with designation R28, C28, S28 and CS28

corrosion resistance can be performed by a comparative analysis of R_p values only, as used and discussed in this work.

Figure 5.3a depicts the evolution of open circuit potentials (OCP), Fig. 5.3b pres-ents the evolution of polarization resistance (R_p) values as recorded via LPR mea-surements, together with the OCP records. Figure 5.3a depicts the OCP evolution for the discussed in this paper series of specimens, cured for 24 h only ("24-hour" groups R, C, S and CS).

The OCP values, recorded for identical, but cured for 28 days specimen groups ("28-day" groups R28, C28, S28 and CS28) are also presented for comparative purposes. The 28-day cured specimens are not subject to this work. However, in order to elucidate the initial electrochemical state in terms of OCP records, and mainly to clarify discrepancies between observed and generally expected response, the 28-day series is briefly discussed with this regard. To be noted is that the OCP values for the 28-day series were recorded also during curing of 1–28 days, when no further conditioning or external factors applied. The OCP values for groups S and CS (stray current involved), in both 24-h and 28-day cured groups, were recorded after current interruption and 24 h decay respectively.

5.3.1.2 OCP Values and Curing Age

In this study, except the conditioning regimes and relevant environment, factors as steel pre-treatment, mortar curing and the additional variable "stray current" are determining the electrochemical response and OCP development, respectively. As can be observed in Fig. 5.3a, the majority of OCP values for the 24-h cured specimens R, C, S and CS until the 28 days of age and beyond fall in the cathodic region and were more negative than the threshold of passivity, i.e., more cathodic than the generally accepted -200 ± 70 mV vs SCE for reinforced mortar (concrete) systems [15]. These OCP values account for active state of the steel reinforcement in all conditions, including the control R group, for at least until 28 days of age. Except the conditioning regimes and external influences (as stray current) there are three main reasons/factors for the observed behavior:

(i) The steel surface properties prior to casting
(ii) The maturity of the cement-based matrix, which determines the pH of the pore solution
(iii) The porosity/permeability of the mortar bulk determining chloride ions penetration in corroding conditions, as well as water penetration and oxygen levels in all conditions

The first factor (steel surface properties) will affect both 24-h and 28-days cured groups. This aspect is of importance in the sense that clean steel surface would be relatively more active compared to oxide layer-covered ("as received") steel, until a stable passive layer is formed in the high pH environment of the mortar pore network. The pH of the pore solution is initially >13.5 in the first hours of cement hydration and stabilizes further to approx. pH of 12.9. In a medium of pH > 13.5 steel is active, following fundamental electrochemical thermodynamics.

The second and third factors, related to cement-based material properties, would predominantly affect the response in the 24-h cured group, whereas different behavior would be expected in the 28-days cured group. This is denoted to a "fresh" state in the former case (only 24 h curing) and higher maturity level in the latter case (the generally employed in cement-based material science 28-day curing). On the other hand, this means that for the period of 1d to 28 days of monitoring, both 24-h and 28-days cured specimen groups should exhibit similar OCP records. However, variation in OCP for the 24-h cured groups will be additionally present as a result from the conditioning regimes in that period. In contrast, the OCP records for the 28-day cured specimens in the same time interval will represent the steel response within cement hydration only.

To this end, Fig. 5.3a can be considered as presenting three main regions of OCP evolution: initial response – *region 1* and day 1, followed by 3–28 days of conditioning – *region 2*, and beyond 28 days to final response at the end of the test for the 24-h cured group – *region 3* (the tests for the 28-day cured series of specimens is ongoing).

5.3.1.3   OCP Evolution Until 28 Days of Age

For all specimens and irrelevant of the curing period (i.e., both 24-h and 28-day cured groups), the initial readings in *region 1*, Fig. 5.3a depict cathodic OCP values in the range of −550 mV to −420 mV, which are far beyond the passivity threshold. This is as expected due to the surface condition of the embedded steel (clean surface), the fresh mortar matrix and related phenomena within cement hydration and pH of the pore solution, as previously discussed. In other words, the electrochemical cleaning of the embedded steel, performed prior to casting, results in a "bare" steel surface, which will be active in alkaline environment of pH > 13.5 until a passive layer is formed and the pH of the environment stabilizes around 12.9.

In *region 2*, Fig. 5.3a, i.e., 3 days until 28 days period for the 24-h cured groups (within conditioning) and the 28-day cured groups (curing only), the OCP records reflect the steel electrochemical response within pore network and steel/cement pate interface development. These are in terms of pore solution chemistry alterations as well as passive layer stabilization (for control cases) or corrosion initiation (for corroding cases). The development of the passive layer and further stabilization is illustrated by the initial fluctuations of OCP values for the control groups R and R28 and stabilization further on, towards more anodic values. For the 28 day-cured control specimens (R28 in Fig. 5.3a), OCP values in the passive domain only were recorded after 28 days of age. In contrast, stabilization of the passive layer for specimens R in the 24-h cured group takes significantly longer period – after 141 days of conditioning, OCPs start tending towards more anodic values. In the former case (28-days group), this development is as expected and already discussed. For the latter case, 24 h curing leads to an open microstructure, non-mature matrix and consequently altered balance of pore water and alkali ions concentration in the pore network. This results in impeding the passive layer formation and development for the control group R, although OCP values are more anodic than these for the corroding groups C and CS between 3 and 28 days of conditioning. For the C and CS groups, active steel surface is expected due to chloride-induced corrosion initiation in both cases and additional stray current contribution in the case of CS specimens.

A pronounced effect of the stray current in group CS towards enhanced corrosion was, however, not observed through OCP records: as shown in Fig. 5.3a, the OCP evolution for specimens C and CS follows similar development over time.

What should be noted is that the stray current in control conditions – group S, was expected to have a negative effect on steel corrosion resistance. This, however, was not observed. On the contrary, the recorded OCP values in *region 2* for specimens S are more anodic than those for specimens R and maintain stability also within region 3. This observation is, however, only relevant for group S which are specimens cured for 24 h only, but is not relevant for groups S28, cured for 28 days. The related phenomena are as follows: for group S, stray current flow through a fresh (non-mature) cement matrix will lead to enhanced water and ion transport due to migration. As a result, cement hydration will be enhanced, leading to a faster development of the pore network and consequently environment, assisting a more rapid stabilization of pore solution and steel/cement paste interface. Previously reported and known are the early stage beneficial effects of stray current on cement-based matrix properties [13].

To this end and with relevance to *region 2* in Fig. 5.3a, the specimens S end up with a superior corrosion resistance, compared to the control specimens R, when 24 h curing only is relevant. In contrast, the stray current applied to group S28 was only after 28 days of curing. Consequently, within *region 2* (3 days until 28 days) the OCP evolution for both R28 and S28 groups is similar, since all specimens in the 28-days cured group can be considered as control, i.e., there are no external factors involved, but only cement hydration is taking place. Therefore, the OCP records for groups R28, C28, CS28 and S28 present similar, although fluctuating, OCP records in *region 2*, which are in the range of these recorded for control specimens R and the specimens S (stray current flow in water environment) from the 24-h cured groups.

5.3.1.4 OCP Records from 28 Days of Age Until the End of the Test (243 days) and R_p Evolution for the 24-h Cured Specimens

A more pronounced differentiation between the 24-h and 28-day cured specimens in electrochemical response, due to environmental effects, stray current and variation of bulk matrix properties, can be already observed in *region 3*, Fig. 5.3a, i.e., from 28 to 243 days. It should be noted that for specimens R28, C28, S28 and CS28, the 28 days' time interval actually corresponds to 28 days of age but 1 day of conditioning only, whereas for specimens R, C, S and CS, the 28 days' time interval corresponds to 28 days of age and 28 days of conditioning. Relevant to *region 3*, time interval of 56 days, it can be observed that for the corroding groups, where NaCl is present in the environment (C, CS, C28 and CS28), all OCP values fall within similar range of values, more cathodic than −550 mV, reflecting active state. Similar to *region 2*, a pronounced effect of the stray current (groups CS) was not distinguished through OCP records, if these are compared to values for specimens C. For the control groups – R and R28, passive layer stabilization and transition from "active" to "passive" domain was observed, especially for group R towards the end of the test. Similarly, the R28 group presents OCP values in the region of passivity, i.e., in the range of -250 mV to -150 mV.

For the 28-day cured groups under stray current, the S28 specimens depict OCPs similar to those for group R28. Apparently, a negative effect of stray current on the corrosion resistance of the embedded steel, especially when applied after 28 days of curing (as is the case for group S28) is not to be observed via OCP records for the relevant time span of this test.

In contrast, for group S, where the stray current application was initiated after 24-h curing only, the initially stable and more anodic values in *region 2* tend to shift to values which are already more cathodic than these for specimens R in the same 24-h cured groups. The OCP development for group S exhibits a stabilization trend below the threshold for active/passive state and remains in the active region. In other words, further records (after 243 days) would possibly reflect already the expected negative effect of stray current on steel corrosion resistance. This hypothesis is supported by the derived R_p values, where a trend towards lower R_p, i.e., lower corrosion resistance, was observed for specimens S, compared to specimens R after prolonged conditioning (Fig. 5.3b). After initially highest R_p values, the corrosion resistance for specimens S was reduced, which is in contrast with the trend of R_p increase for specimens R towards the end of the test.

The R_p values for the corroding, NaCl conditioned, groups C and CS maintain low R_p values throughout the reported period. Lower R_p values were recorded for specimens C if compared to those for specimens CS. However, towards the end of the test the corrosion activity for C and CS is almost equally high. This observation reflects the previously discussed hypothesis for the expected negative effect of stray current in specimens CS only with prolonged conditioning. Obviously, competitive mechanisms act in specimens CS, where on the one hand the stray current has positive effects on bulk matrix properties, similar to specimen S, at early stages. On the other hand, stray current influences chloride ions migration, leading to chloride-induced corrosion and active state, similarly to specimens C. Here again, a conclusive statement on the effect of stray current on corrosion resistance in both S and CS groups cannot be made, if considerations are based only on OCP and R_p evolution during the reported time interval of this experiment.

5.3.2   EIS and PDP Response

EIS is a non-destructive electrochemical technique, which when applied to a reinforced concrete system provides qualitative and quantitative information for both the electrochemical state of the steel reinforcement and the properties of the bulk matrix. EIS was employed in a manner, similar to previously reported such for an otherwise in-depth characterization of steel-concrete systems [17–19]. The results and discussion in this work refer to time intervals within the period of 243 days of conditioning for all studied specimen groups. Both qualification, by simply evaluating and comparing EIS response, and quantification of general corrosion state (deriving R_p values) were performed. Absolute values are not claimed but rather a comparison between equally handled specimens is relevant and discussed.

The EIS response was recorded in the frequency range of 50 kHz to 10 mHz. This frequency range provides information for the contribution of the bulk matrix (high frequency (HF) range > 1 kHz) to the electrochemical response of the embedded steel (middle to low frequency ranges, i.e., < 1 kHz and mainly 1 Hz to 10 mHz). Absolute values for concrete bulk properties can be derived at high frequency range, above the 1 MHz region, e.g., starting at 10 MHz [20]. Therefore, the qualification and quantification of the bulk matrix in this work refers to contribution only of the mortar bulk matrix, in terms of: solid phase, connected and disconnected pore network, together with external medium (solution) contribution, which is negligible if compared to that of the bulk matrix.

5.3.2.1 General Considerations Towards EIS Data Interpretation

If a prompt evaluation of the corrosion state of steel embedded in a cement-based material is aimed at, together with a simplified assessment of the electrical properties of the bulk matrix itself, qualification of the EIS response is a very useful approach. This is especially the case if specimens, conditioned in different environment and hence, expected to present significant variation in properties and response, are subject to investigation. For instance, a reinforced mortar specimen conditioned in NaCl will logically perform different in time, if compared to a control specimen conditioned in water. This is due to the expected chloride-induced steel corrosion in the former case and stabilization of the passive state of the steel reinforcement in the latter case.

Additionally, alterations in the electrical properties of the bulk cement-based matrix, e.g., increased electrical resistivity over time, would be expected in all conditions, due to the process of cement hydration and subsequent matrix densification. Similarly, factors as chemical composition of the external environment, ions and water penetration into the bulk matrix, variation in bulk matrix diffusivity, pore interconnectivity, etc. will determine changes in the electrical properties of the bulk matrix over time. These again will be reflected by the EIS response and will vary not only as a result of the corrosion state of the embedded steel but also as a result of conditioning and treatment. All these features in an experimental EIS response are well visible and can be compared qualitatively for systems as in this work – corroding and control reinforced mortar specimens. Additionally, specimens as the hereby discussed "under current regimes" can be indirectly evaluated by a comparison with the corroding and control specimens.

To this end, the EIS response for the hereby studied R, C, S and CS groups of specimens was qualitatively evaluated first. This approach is mainly relevant to the high frequency domain and bulk matrix response, respectively. Next, quantification of the experimental EIS data was performed and the four specimen groups in this

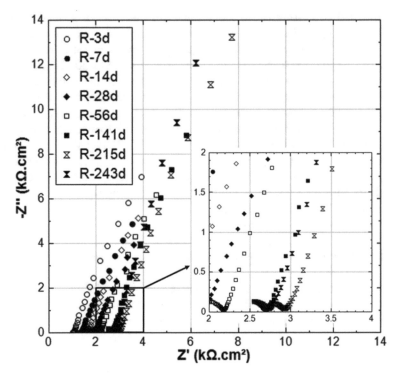

Fig. 5.4 EIS response for Group R as an overlay of 3–243 days

work were compared in view of global corrosion resistance of the embedded steel reinforcement.

The experimental impedance responses, in a Nyquist plot format, for all groups of specimens are presented in Figs. 5.4, 5.5, 5.6 and 5.7 as an overlay from 3 to 243 days per specimens group.

With regard to Group R and S, the shape of the experimental curves in Figs. 5.4 and 5.5 reflects the typical response of steel in a chlorides-free cement-based environment (alkaline medium). The response depicts curves inclined to the imaginary y-axis, reflecting a capacitive-like behavior, or passive state, of the steel reinforcement. This behavior is relevant for group R throughout the test and is more significantly pronounced after 56 days of conditioning, denoted to stabilization of the passive layer over time. The observation is in line with the OCP ennoblement for group R towards the end of the test (Fig. 5.3). For specimens S, although similar to specimens R behavior is relevant at initial time intervals – 3 to 28 days, the magnitude of impedance tends to lower values towards the end of the test, reflected by an inclination of the EIS response towards the real, x-axis. This is also in line with the more cathodic OCP values, recorded for group S towards the end of the test (Fig. 5.3).

In contrast to R and S groups, the EIS responses for groups C and CS, Figs. 5.6 and 5.7, show a clear evidence of active corrosion on the steel surface. Starting at

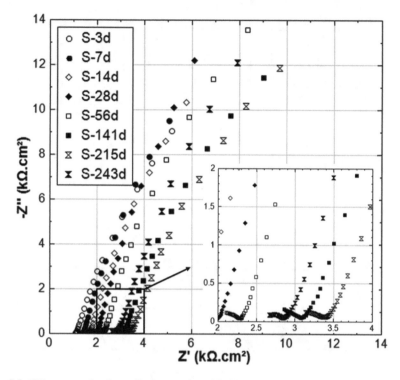

Fig. 5.5 EIS response for Group S as an overlay of 3–243 days

very early stage (3 days) the response is already inclined to the real axis semicircle with a decreasing magnitude of impedance |Z| towards the 243 days. The shape of this EIS response was also largely reported to be due to the presence of chloride ions on the steel surface and increasingly active corrosion state [21].

The Nyquist plots in Figs. 5.4, 5.5, 5.6 and 5.7 also reflect the changes of EIS response in view of bulk matrix characteristics – this is the response in the high frequency domain (inlets in Figs. 5.4, 5.5, 5.6, and 5.7). Additionally and for a more clear comparison in the HF domain, Figs. 5.8 and 5.9 (marked regions) present the EIS response as an overly of magnitude of |Z| in Bode plots format. As previously mentioned, the HF response in this work is with regard contribution only of the bulk matrix, since the measurements were performed, starting from 50 kHz. Therefore a qualitative evaluation and comparison was only made in this frequency domain and is presented in what follows.

5.3.2.2   High Frequency Response and Bulk Matrix Properties

The inlets in Figs. 5.4, 5.5, 5.6 and 5.7 present a more detailed view of the HF response. In other words, the bulk matrix contribution is reflected by the initial, semicircular portion of the EIS experimental curves as depicted in the Nyquist plots.

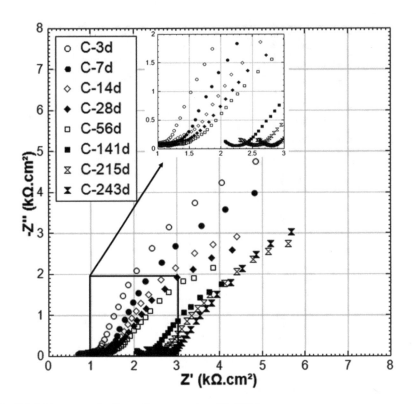

Fig. 5.6 EIS response for Group C as an overlay of 3–243 days

As seen from the plots, an increase in bulk matrix resistance (increase of real |Z| values in the HF domain over time) is relevant for all specimen groups within treatment, irrespective of the external environment. The experimental curves in Figs. 5.4, 5.5, 5.6 and 5.7 are also depicted in a Bode format, Figs. 5.8 and 5.9, presenting log of impedance |Z| vs log frequency. This presentation gives a clearer overview of the changes in the HF domain over time (marked regions in Figs. 5.8 and 5.9), where a trend towards increase of bulk matrix resistance can be observed for all groups.

The HF EIS response, Fig. 5.8, accounts for initially similar properties of the bulk matrix for specimens R and S. Later on and towards the end of the test, bulk matrix densification and/or reduced permeability of the pore network would be relevant for specimens S, evident from increased magnitude of |Z| in the HF domain. The increase of bulk matrix resistance was as expected and due to cement hydration. The result is filling-up of "empty" (pore) space with hydration products, which leads to microstructural development and therefore reduced porosity and permeability. Since cement hydration is affected by water and ions transport/interaction within the pore network, it is also logic that these processes will be enhanced in conditions of current flow, if compared to control conditions. In other words, in addition to diffusion from concentration gradients, ion and water migration take place in specimens

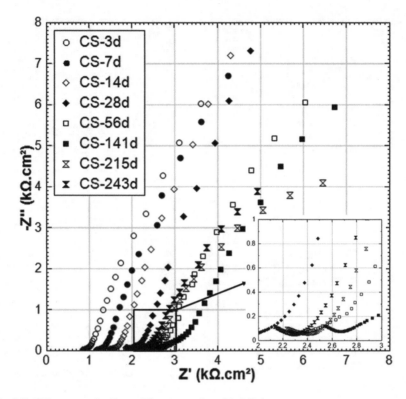

Fig. 5.7 EIS response for Group CS as an overlay of 3–243 days

S, which in turn enhances the related cement hydration mechanisms. The result will be higher bulk matrix resistance as actually recorded by the EIS response, Fig. 5.8.

Different behavior was observed for groups C and CS, additionally related to the chloride ions in the external environment on the one hand (specimens C and CS) and the effect of chloride ions and current flow on the other hand (specimens CS only). Well known is the effect of NaCl on cement hydration. This is in view of acceleration of the hydration process, resulting in matrix densification [22]. This process would be expected to be more pronounced within the additional effect of ions and water migration when current flow is involved, as in CS specimens. The EIS response reflects these changes – Fig. 5.9. The initial HF response for specimens CS showed higher HF impedance values (starting at 3, 7 and 14 days of treatment), compared to that for specimens C at the same time intervals. Later on and towards the end of the test, e.g., 141–243 days, the HF impedance reflecting bulk matrix contribution for both C and CS specimens shows a similar trend and range, Fig. 5.9.

What can be concluded from evaluation and qualitative interpretation of the HF response is that stray current affects the bulk matrix properties. The effect is positive in view of increased bulk matrix resistance and is more pronounced at later stages

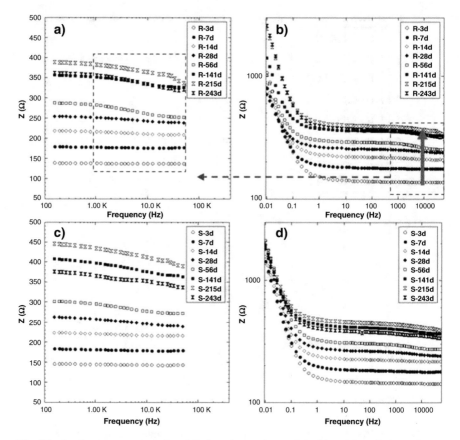

Fig. 5.8 Overlay response from 3 to 243 days for specimens R (**a**, **b**) and S (**c**, **d**) in Bode plot format – overlay magnitude |Z| in HF-MF domain (50 kHz – 150 Hz)

(after 56 days) for specimens, treated in water (group S compared to group R, Fig. 5.8) and at earlier stages – 3 to 28 days for specimens, conditioned in NaCl (specimens C and CS, Fig. 5.9).

5.3.2.3 Quantification of EIS Response and Global Corrosion State, Including PDP Test

Quantitative information from the EIS response is normally obtained by fitting the experimental data, using the relevant electrochemical software, with an equivalent electrical circuit. The electrical circuit is a sequence of electrical parameters, in series with the electrolyte (external medium) resistance. Each parameter, or a combination thereof, represents a physical meaning, relevant to the EIS response of the system under study. The theory and practice behind EIS response electrochemical

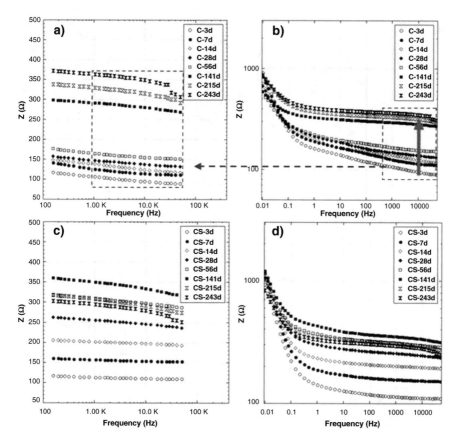

Fig. 5.9 Overlay response from 3 to 243 days for specimens C (**a**, **b**) and CS (**c**, **d**) in Bode plot format – overlay magnitude |Z| in HF-MF domain (50 kHz – 150 Hz)

fit and simulation, in general or for a reinforced cement-based system, are not subject to this work and can be found in related state of the art [23–25].

An example for experimental response and fit for the studied specimens in this work is presented in Fig. 5.10 in both Nyquist and Bode plots format. The response is not normalized in the examples of Fig. 5.10, since equal geometry of the specimens and steel surface area were relevant for all cases. Figure 5.10a depicts the initial EIS response, after 3 days of treatment, and fit for specimen S. Figure 5.10b depicts the EIS response and fit for a specimen from group CS after 215 days of treatment. Two types of equivalent electrical circuits were used (inlets in Nyquist plots in Fig. 5.10): for the cases of specimens R and S (chloride-free environment) the EIS data were fitted by a circuit with two hierarchical time constants in series with the electrolyte resistance. For the specimens in groups C and CS (chloride-containing external medium), the circuit was composed of three time constant.

For both circuits, R_0 is the electrolyte resistance, together with the contribution of the mortar bulk. In the two-time constant circuit, the first time constant (R_1 and

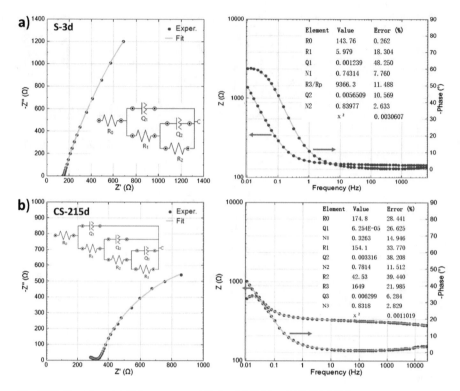

Fig. 5.10 Experimental EIS response and fit for: (**a**) S-3d and (**b**) CS-215d

Q_1) is attributed to the pore network of the mortar matrix, where R_1 and Q_1 are the resistance and pseudo-capacitance of the bulk matrix (the replacement of pure capacitance C with constant phase element Q is a general approach when higher level of heterogeneity is relevant to the system under study [18]). The second time constant for the two-time constant circuit (R_2 and Q_2) deals with the electrochemical reaction (charge transfer process and mass transport process) on the steel surface. The resistance R_2 in this case represents the polarization resistance R_p. In the three-time constant circuit, R_3 and Q_3 correspond to the electrochemical reaction on the steel surface. The first two time constants are attributed to the bulk matrix + electrolyte resistance, including disconnected pore space (R_1 and Q_1), while R_2 and Q_2 represent the resistance and pseudo-capacitance of the connected pore space. A separation between connected and disconnected pore space in specimens C and CS (i.e., additional time constant for the HF domain) is more pronounced and relevant as a result from the influence of the chloride-containing medium on the bulk matrix properties and subsequent alterations in cement hydration and chloride binding mechanisms [22].

The best fit parameters after electrochemical fit and simulation of the relevant system are also included in Fig. 5.10 – inlet in the Bode plots. Detailed presentation

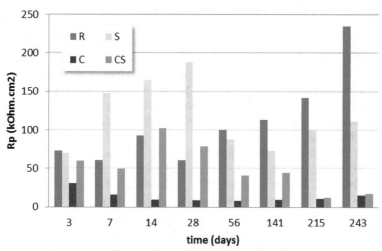

Fig. 5.11 Evolution of polarizations resistance (R_p) for control group (R), stray current group (S), corroding group (C) and corrosion under stray current group (CS). R_p values are as derived from the best fit paramaters from fitting the EIS response, i.e., determined from the evolution of recorded resistance for the last time constant of the employed equivalen circuits

and discussion of all derived parameters is not subject to this work, since the purpose of EIS presentation and discussion was to mainly illustrate the application of EIS to the system under study and correlate results to those from DC electrochemical techniques and derived parameters (Fig. 5.3). Therefore, Fig. 5.11 presents the evolution of global corrosion resistance for all tested specimen groups with time of conditioning (from 3 to 243 days), in terms of polarization resistance values (R_p) as derived from EIS tests.

As can be observed in Fig. 5.11, the R_p values for the control specimens R (non-corroding, water environment) and the specimens S (under stray current in water environment) show a generally increasing trend from 3 to 28 days of treatment. This is in accordance with the expected stabilization of the passive film in the former case (specimens R), although fluctuations in R_p values were recorded through EIS. For the latter case (specimens S), the influence of stray current was expected to potentially exert negative effect on the passive layer formation and stabilization.

However, this was not as observed. On the contrary, the R_p values for specimens S gradually increase in the period of 3–28 days and maintain the highest values among all tested conditions. The reason for this performance for specimens R and S is the already discussed synergetic effect of electrochemically cleaned, i.e., initially active surface and the effects of fresh (non-mature) cement matrix, together with concentration gradients of external environment – cement-based material and steel/cement paste interface. For specimens R all these result in a delay in passive layer formation and stabilization, whereas for specimens S, the current flow induces a positive effect of enhanced cement hydration and consequently favorable environment for the steel reinforcement. While for specimens S the trend of R_p evolution as derived from EIS is very well in line with the R_p values derived from LPR (Fig. 5.3), i.e., increasing corrosion resistance between 3 and 28 days, this is not entirely the case for specimens R, where R_p values from EIS and LPR significantly differ. This observation supports the hypothesis for the initially more rapid formation and stabilization of the product layer on the steel surface of specimens S, if compared to specimens R for that time period of the test. This results in a higher R_p due to the contribution of a product layer with higher resistance in specimen S, if compared to specimens R, where the derived R_p (from both LPR and EIS) is mainly related to charge transfer resistance.

It should be noted that the absolute values of R_p as derived from both methods are not entirely comparable due to the fact that LRP is a DC and EIS – an AC measurement. Although LPR induces a small DC polarization, EIS applies a 10 mV AC perturbation only. In the case of not yet stable passive (or product) layer on the steel surface, and in view of the range of recorded OCP values (Fig. 5.3), the LPR measurement can result in under- or overestimation of R_p. Therefore, a comparison of trends is always relevant, rather than discussion of absolute values. Additionally, the EIS tests exert minimum or none effect on the forming passive layer on the steel surface and include the resistance of the undisturbed product layer at the time of measurement, in addition to the charge transfer resistance. This gives the global R_p values within EIS tests, which can end up higher if compared to R_p, derived from LPR measurement in these relevant conditions.

From 28 days onwards and until the end of the test of 243 days of conditioning, the already stabilized steel surface and the increased maturity level of the cement-based matrix result in an increasing trend only for the R_p values for specimens R, derived from both EIS (Fig. 5.11) and LPR (Fig. 5.3). In fact, at the end of the test, the R_p values and corrosion resistance, respectively, for specimens R are the highest from all tested cases (as expected). In contrast, the decreasing trend of R_p for specimens S after 28 days and until 243 days already proves the negative effect of stray current on steel passivity, although the corrosion state of specimens S is still superior if compared to the corroding specimens C and CS. In other words, stray current, even in conditions of chloride-free environment, would potentially result in significantly reduced corrosion resistance in the long term.

The R_p values derived from EIS for specimens C and CS (chloride-containing environment) are significantly lower, compared to these for groups R and S, which was as expected. Both corroding groups present more significant fluctuations in R_p values between 3 and 28 days, due to reasons already discussed for groups R and S. Additionally, the combined action of chloride ions and stray current in group CS results in initially higher R_p compared to group C, with a decreasing trend towards the end of the test. The evolution of R_p derived from EIS is in line with that derived from LPR and the responsible phenomena are as already discussed with respect to Fig. 5.3. Additionally, the reason behind similar EIS and LPR response for the corroding groups at all time intervals (which is in contrast to that for the groups S and R for the initial period) is because of the predominance of corrosion initiation and propagation in these two cases, rather than processes related to product layer

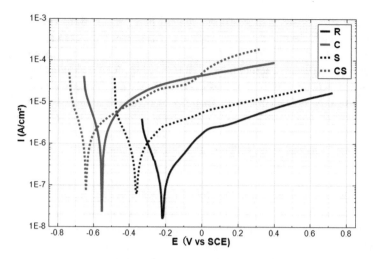

Fig. 5.12 Potentio-dynamic curves for all specimens at 243 days of age

stabilization. Therefore, both techniques result in determination of similarly active state for specimens C and CS.

If the significance of stray current within corrosion initiation and propagation in chloride-containing environment is to be evaluated (specimens C and CS), both EIS and LPR show that a clear evidence for a substantial stray current effect is not present. The corrosion resistance for both C and CS groups was similar at the end of the test. This is most likely denoted to the fact that within the period of 3–28 days, the stray current induced positive effects in view of ion and water migration and enhanced cement hydration – also evident from the response for both C and CS in the high frequency domain of EIS measurements (Fig. 5.9). Therefore, in specimens CS, competing mechanisms were initially involved, and the predominance of chloride-induced corrosion acceleration due to stray current is only relevant on later stages of cement hydration and treatment. These considerations are actually supported by the results from potentio-dynamic polarization, performed at the end of the test – Fig. 5.12.

Figure 5.12 presents the polarization curves for all investigated specimens at 243 days of age. The measurements were performed after 24 h depolarization for the under current regimes (CS and S), i.e., after awaiting for the establishment of a stable OCP of the steel reinforcement. What can be observed is that the most noble potential (approx. -220 mV) and lowest corrosion current were recorded for specimen R – as expected and in line with all other tests. The corrosion potential for specimens S is more cathodic (approx. -380 mV) and the corrosion current is approximately one order higher than that for specimens R. Additionally, larger anodic current within polarization was recorded for specimen S. Here again, stray current induces corrosion in reinforced cement-based materials and reduces the corrosion resistance, despite the chloride-free environment for specimens S.

The cathodic shift of corrosion potential is more evident for the specimens C and CS, both already more cathodic than −500 mV, Fig. 5.12. Additionally, the corrosion current densities for both C and CS are already significantly higher. These are in line with the, as recorded from all other tests, active state of specimens C and CS. If a comparison is made between specimens C and CS, what can be observed is that, although specimens CS present a more cathodic corrosion potential, the corrosion current is in the same order as that for specimens C. This means that both groups are relatively similar in corrosion activity, as also derived from EIS and LPR in view of similar R_p values at the end of the test. However, the initially higher corrosion resistance in specimens CS (Figs. 5.3 and 5.7) and the potentially positive effect of the stray current in view of ion transport and enhanced cement hydration, hence favorable environment at the steel surface, are well reflected by the behavior of CS with external polarization. As can be observed in Fig. 5.12, the lower anodic currents with external polarization and increase only after the region of around 0 V are denoted to impeded dissolution, i.e., to a more resistive or larger in surface area product layer on the steel surface in CS, if compared to specimen C. This results in initially higher R_p values for specimen CS (as recorded and already discussed), despite the otherwise similarly to C active state.

What can be concluded is that the effect of stray current for both chloride-free (specimens S) and chloride-containing (specimens CS) conditions is predominantly positive in the initial stages of this test and exerts the expected negative influence towards corrosion acceleration after a prolonged treatment and within already a more stable maturity level of the cement-based matrix. This also means that the effect of the cement-based material in reinforced cement-based system is of significant importance and largely determines the electrochemical state of the steel reinforcement.

5.4   Conclusions

In this work, a comparative study was proposed for investigating the different effects of stray current on bulk matrix and the corrosion behavior of embedded steel. Based on the experimental and analytical results, the following conclusions can be drawn:

1. The effect of stray current on concrete bulk matrix properties, together with steel corrosion response, is significantly determined by the external environment, as well as by the level of maturity of the cement-based bulk matrix. Stray current is predominantly positive in the initial stages of this test, but the expected negative influence towards corrosion acceleration was observed after a prolonged treatment, when a stable maturity level of the cement-based matrix was at hand.
2. For chloride-free environment, the effect of the chosen stray current level was not significant, although lower corrosion resistance of the steel rebar was

recorded after longer exposure of 240 days. Positive effects of the stray current were observed in the early stages of the experiment (until 28 days), which were related to enhanced ion and water migration and consequently increased cement hydration and passive film development in the highly alkaline environment of the mortar bulk matrix.

3. In terms of the chloride-contained environment (C and CS groups), active steel surface was detected as expected, due to chloride-induced corrosion initiation in both cases and additional stray current contribution in the case of CS specimens. However, the pronounced effect of the stray current in group CS towards enhanced corrosion was not observed. Based on this, a competitive mechanisms acting in specimens CS can be proposed: on the one hand the stray current has positive effects on bulk matrix properties, similar to specimen S, at early stages. On the other hand, stray current influences chloride ions migration, leading to chloride-induced corrosion and active state, similarly to specimens C.

References

1. Bertolini L, Carsana M, Pedeferri P (2007) Corrosion behaviour of steel in concrete in the presence of stray current. Corros Sci 49(3):1056–1068. doi:10.1016/j.corsci.2006.05.048
2. Metwally IA, Al-Mandhari HM, Gastli A, Al-Bimani A (2008) Stray currents of ESP well casings. Eng Anal Bound Elem 32(1):32–40. doi:10.1016/j.enganabound.2007.06.003
3. Revie RW (2008) Corrosion and corrosion control. Wiley, New York
4. Wang CL, Ma CY, Wang Z (2007) Analysis of stray current in metro DC tractionpower system. Urban Mass Transit 3:51–56
5. Wu M (2011) Progress of the durability of concrete structure of the subway (trans: Hainan University CoCE, Architecture, Guizhou University CoC, Architecture E, Hainan Society of T, Applied M). In: 1st international conference on civil engineering, architecture and building materials, CEABM 2011, .vol 250–253. Haikou. doi:10.4028/www.scientific.net/AMR.250-253.1456
6. Sandrolini L (2013) Analysis of the insulation resistances of a high-speed rail transit system viaduct for the assessment of stray current interference. Part 1: measurement. Electr Power Syst Res 103:241–247. doi:10.1016/j.epsr.2013.04.011
7. Chen Z, Koleva D, Koenders E, van Breugel K (2015) Stray current induced corrosion control in reinforced concrete by addition of carbon fiber and silica fume. MRS Online Proceedings Library 1768:null-null. doi:10.1557/opl.2015.320
8. Radeka R, Zorovic D, Barisin D (1980) Influence of frequency of alternating current on corrosion of steel in seawater. Anti-corros Methods Mater 27(4):13–15+19
9. Kolar V, Hrbac R (2014) Measurement of ground currents leaking from DC electric traction. In: Electric Power Engineering (EPE), Proceedings of the 2014 15th international scientific conference on, 2014. IEEE, pp 613–617
10. Tullmin M (2003–2007). http://www.corrosion-club.com/stfeature1.htm
11. Chang JJ (2003) Bond degradation due to the desalination process. Constr Build Mater 17(4):281–287. doi:10.1016/s0950-0618(02)00113-7
12. Chang JJ, Yeih W, Huang R (1999) Degradation of the bond strength between rebar and concrete due to the impressed cathodic current. J Mar Sci Technol 7(2):89–93
13. Susanto A, Koleva DA, Copuroglu O, van Beek K, van Breugel K (2013) Mechanical, electrical and microstructural properties of cement-based materials in conditions of stray current flow. J Adv Concr Technol 11(3):119–134. doi:10.3151/jact.11.119

14. Standard practice for preparing, cleaning, and evaluating corrosion test specimens (2003) Annual book of ASTM standards (2):17–25
15. Alonso C, Castellote M, Andrade C (2002) Chloride threshold dependence of pitting potential of reinforcements. Electrochim Acta 47(21):3469–3481. doi:10.1016/S0013-4686(02)00283-9
16. Alghamdi SA, Ahmad S (2014) Service life prediction of RC structures based on correlation between electrochemical and gravimetric reinforcement corrosion rates. Cem Concr Compos 47:64–68. doi:10.1016/j.cemconcomp.2013.06.003
17. Sagüés AA, Pech-Canul MA, Shahid Al-Mansur AKM (2002) Corrosion macrocell behavior of reinforcing steel in partially submerged concrete columns. Corros Sci 45(1):7–32. doi:10.1016/S0010-938X(02)00087-2
18. Feliu V, González JA, Feliu S (2004) Algorithm for extracting corrosion parameters from the response of the steel-concrete system to a current pulse. J Electrochem Soc 151(3):B134–B140. doi:10.1149/1.1643737
19. Koleva DA, Van Breugel K, De W, Delft TU, Civil E, Geosciences, Tu Delft DUoT (2007) Corrosion and protection in reinforced concrete: pulse cathodic protection: an improved cost-effective alternative
20. Keddam M, Takenouti H, Nóvoa XR, Andrade C, Alonso C (1997) Impedance measurements on cement paste. Cement Con Res 27(8):1191–1201. doi:http://dx.doi.org/10.1016/S0008-8846(97)00117-8
21. Koleva DA, van Breugel K, de Wit JHW, van Westing E, Boshkov N, Fraaij ALA (2007) Electrochemical behavior, microstructural analysis, and morphological observations in reinforced mortar subjected to chloride ingress. J Electrochem Soc 154(3):E45. doi:10.1149/1.2431318
22. Florea MVA, Brouwers HJH (2012) Chloride binding related to hydration products: part I: ordinary Portland cement. Cem Concr Res 42(2):282–290. doi:10.1016/j.cemconres.2011.09.016. 10.1016/j.conbuildmat.2011.07.045; Brouwers HJH (2011) A hydration model of Portland cement using the work of powers and brownyard, http://www.cement.org, Eindhoven, University of Technology & Portland Cement Association. ISBN: 978-90-6814-184-9
23. Andrade C, Soler L, Novoa XR (1995) Advances in electrochemical impedance measurements in reinforced concrete. Mater Sci Forum 192-194(pt 2):843–856
24. Cui W, Shi Z, Song G, Lin H, Cao CN (1998) Electrochemical study on the reinforced concrete during curing. Corros Sci Prot Technol 10(4):201–207
25. Wenger F, Galland J (1990) Analysis of local corrosion of large metallic structures or reinforced concrete structures by electrochemical impedance spectroscopy (EIS). Electrochim Acta 35(10):1573–1578. doi:10.1016/0013-4686(90)80012-D

Chapter 6
The Effect of Nitrogen-Doped Mesoporous Carbon Spheres (NMCSs) on the Electrochemical Behavior of Carbon Steel in Simulated Concrete Pore Water

H. Mahmoud, J. Tang, Dessi A. Koleva, J. Liu, Y. Yamauchi, and M. Tade

Abstract The influence of highly nitrogen-doped mesoporous carbon spheres (NMCSs) (internal pore size of 5.4–16 nm) on the electrochemical response of low carbon steel (St37) in model alkaline solutions of pH 13.9 and 12.8 was studied, using Open Circuit Potential (OCP) monitoring, Electrochemical Impedance Spectroscopy (EIS) and Cyclic Voltammetry (CV). Prior to adding the NMCSs in the relevant solutions, they were characterized in the same model medium by measuring their Zeta-potential, hydrodynamic radius and particle size distribution, using dynamic light scattering (DLS) and transmission electron microscopy (TEM). In alkaline environment of pH 13.9 and 12.8, which simulates the concrete pore water of fresh and mature concrete, the DLS measurements indicated that the hydrodynamic radius of NMCSs particle varied from 296 nm to 183, respectively. According to the Zeta-potential measurements in the same solutions, the NMCSs were slightly positively charged.

H. Mahmoud (✉) • D.A. Koleva
Faculty of Civil Engineering and Geosciences, Delft University of Technology,
Section of Materials and Environment, Stevinweg 1, 2628 CN Delft, The Netherlands
e-mail: H.AminHassan@tudelft.nl

J. Tang • Y. Yamauchi
National Institute for Materials Science (NIMS),
1-1 Namiki, Tsukuba, Ibaraki 305-0044, Japan

J. Liu • M. Tade
Curtin University of Technology, Faculty of Science and Engineering, Department of
Chemical & Petroleum Engineering, GPO Box U1987, Perth, WA, Australia

© Springer International Publishing AG 2017   109
L.E. Rendon Diaz Miron, D.A. Koleva (eds.), *Concrete Durability*,
DOI 10.1007/978-3-319-55463-1_6

The addition of 0.016 wt.% of NMCSs to the model medium induced certain variation in the electrochemical response of the tested steel. In alkaline solutions of pH 12.8, the presence of NMCSs in the passive film/solution interface induced a delay in the formation of a stable passive film. On the other hand, in solutions of pH 13.9, the higher corrosion activity on the steel surface, enhanced by high pH, was limited by adsorption of NMCSs on the film/substrate interface. In addition competing mechanisms of active state, i.e., enhanced oxidation on the one hand, and particles adsorption on anodic sites and oxidation limitation, on the other hand, was relevant in solution of pH 13.9 inducing larger fluctuations in impedance response and stabilization only towards the end of the testing period.

Except steel electrochemical response, the properties of the cement-based bulk matrix were investigated in the presence of the aforementioned additives. The mortar bulk matrix properties were highly affected by NMCSs. The lowest electrical resistivity values were recorded in mortar specimens with mixed-in 0.025 wt.% NMCSs (with respect to dry cement weight). Furthermore, the addition of 0.025 wt.% NMCSs increased the compressive strength when compared to control specimens. The presence of F127 as a dispersing agent for NMCSs was found to be not suitable for reinforced concrete applications, which is in view of the reduced mechanical strength and electrical resistivity of the cement-based bulk matrix. This is in addition to the adverse effect on the formation of electrochemically stable passive layer on steel surface in alkaline medium in its presence.

Keywords NMCSs • Steel corrosion • Alkaline solutions • Cement-based materials

6.1   General Introduction

Concrete is the most widely used construction material worldwide. It is considered to be a durable material; however, premature deterioration of reinforced concrete structures (RCS) can be induced by chloride ingress and/or carbonation [1–3]. The result is subsequent steel corrosion initiation and propagation, which significantly affect the RCS durability. Therefore, there is a constant drive to improve the properties of RCS, e.g., by tailoring tensile or flexural mechanical properties in order to achieve superior damage residence and increased RCS' durability [4, 5].

Corrosion of the steel reinforcement is widely recognized to be the main cause for RCS durability loss [6–8], reflected by the enormous economic consequences from maintenance and repair [3]. Over the last decade there has been a rapid growth in the diversity of applied methods to improve RCS durability by minimizing or preventing corrosion of the steel reinforcement.

The state-of-the-art reports on numerous corrosion protection methods and approaches [9] such as: the introduction of alternative high corrosion resistant reinforcements [10], improving the concrete barrier properties [11–13], cathodic protection [14, 15], corrosion inhibitors [16], electrochemical chloride extraction [15], protective coatings [17], etc. All these are meant to meet given requirements of serviceability, strength and stability throughout the designed service life of a reinforced concrete structure, without significant loss of utility or excessive unforeseen maintenance [18].

Many studies have been focused on the performance of cement-based composite materials incorporating allotropes of carbon, e.g., carbon nanotubes (CNTs), carbon nanofibers (CNFs) and graphene oxide, all of these generally used to improve the mechanical performance of cementitious materials [11, 12, 19]. However, the complete understanding of the mechanisms upon which these additions react with the cement-based materials and affect further the corrosion performance of the embedded steel is still a challenge.

In the frame of a recent approach to corrosion control in reinforced concrete using various nanomaterials, admixed within the cement-based composite, the possible application of nitrogen-doped mesoporous carbon spheres (NMCSs) was also considered. Keeping in mind the high surface area of NMCSs, including the expected alteration of electrical properties of the bulk material in their presence, several hypothetical applications were evaluated. For example, the NMCSs-modified cement-based bulk matrix would theoretically present a reduced electrical resistance. Hence, possible application for NMCSs in cement-based layers within cathodic protection systems, or control of stray current-induced corrosion in RCS, could be a novel and feasible approach.

The NMCSs showed high chemical inertness, high surface area (from 343 to 363 $m^2 \ g^{-1}$) and large porosity (from 0.45 to 0.48 $cm^3 \ g^{-1}$), good electrical conductivity and elevated electro-catalytic properties especially for oxygen reduction reactions [20–22]. With respect to the application of NMCSs in RCS, it should be noted that, within admixing of the NMCSs in cement-based composite materials, their dispersion in the bulk matrix is expected to be a critical parameter. This holds for carbon-based materials (as admixtures) in general and is expected to strongly influence the properties of the final cement-based composite [23]. One of the reasons for a potentially poor dispersion may be the cement particles agglomeration [24] and/or the tendency of carbon particles to agglomerate themselves due to the possible interference of attraction (Van der Waals) forces [25]. Different dispersion methods have been employed and reported for both aqueous (water and solutions) and solid (cement-based) medium [20, 23–26]. For example, in order to achieve homogenous dispersion of carbon nanomaterials in water, the use of various surfactants is generally employed [27, 28], together with or without ultrasonication [29]. In the present work, Pluronic F127 was used as a dispersing agent for NMCSs in the aqueous model medium [21, 30], while either ultrasonication or addition only of F127 was used for the solid (cement-based) environment.

Regarding the incorporation of NMCSs in reinforced concrete, certain effects on steel corrosion resistance, together with possibly superior concrete material properties, are expected. For example, improving mechanical properties, decreasing electrical resistivity, alteration in the bulk matrix properties and microstructure in terms

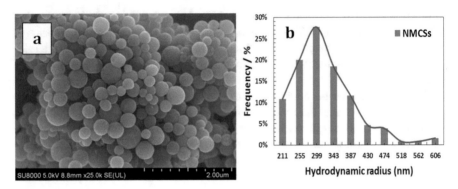

Fig. 6.1 Shows **a**) TEM image, and **b**) hydrodynamic radius distribution of as-recieved NMCSs particles

reduced porosity/permeability, etc. could be achieved. These in turn are also expected to lead to an increase of steel corrosion resistance.

This work briefly introduces results from tests on global cement-based material properties (mechanical and electrical properties of mortar) in the presence of 0.016 wt.% NMCSs. Next, the study aims to evaluate the effect of NMCSs on the electrochemical behavior of steel in model solution as an approach that normally precedes tests in reinforced mortar and concrete. Model solutions of NaOH were used, rather than complex simulated environment (as cement extract for example), in order to allow direct correlation to reported results for electrochemical performance of steel in general, when both alkaline environment and carbon-based additives are concerned. Next, the model solution of pH 13.9 referred to simulating concrete pore water at early stages (in the range of hours) of cement hydration, whereas the solution of pH 12.8 simulated the concrete pore water at later stages of cement hydration.

6.2 Experimental Materials and Methods

6.2.1 NMCSs Preparation and Characterization

NMCSs were obtained through self-polymerization of dopamine (DA) and spontaneous co-assembly of PS-b-PEO di-block copolymer micelles, forming PDA/PS$_{173}$-b-PEO$_{170}$ composite spheres. The micelles of high molecular weight block polymer PS$_{173}$-b-PEO$_{170}$ acted as a sacrificial pore-forming agent and were subsequently removed during a carbonization process, leaving mesopores in the range of 5–16 nm in the carbon spheres [20].

Figure 6.1 shows a TEM image and hydrodynamic radius distribution of the as received of the NMCSs, as obtained after carbonization of the PDA/PS-b-PEO at 800 °C. All NMCSs showed relatively rough surface, however, presenting uniformly distributed surface features with an average sphere size varying from approx.

Table 6.1 Mortar specimen's designation

Mortar specimens	NMCS/ wt.% (Cement wt.)	F127/ wt.% (mix water wt.)	Ultrasonification/ min	No. of samples
Blank	0	0	0	4
NMCS	0.025	0	5	4
NMCS + F127	0.025	0.1	0	4
F127	0	0.1	0	4

300 to 400 nm, Fig. 6.1 (particles with a minimum size of 187 nm and a maximum size of 610 nm were observed).

6.2.2   Cement-Based Materials

Preliminary studies on the effect of NMCSs on the global performance of cement-based materials were performed in order to account for their hypothesized influence in reinforced concrete systems in view of bulk matrix properties. Compressive strength tests and electrical resistivity measurements were performed on 40 × 40 × 40 mm mortar cubes with admixed 0.025 wt.% NMCSs (with respect to dry cement weight). The mortar specimens were cast using ordinary Portland cement (OPC) CEM I 42.5 N (ENCI, NL), a water-to-cement ratio (w/c) of 0.5 and 1:3 cement-to-sand ratio. The NMCSs were added within mortar casting as NMCSs only or as a mixture of NMCSs + Pluronic F127, where F127 was employed as a dispersing agent. In the former case (NMCSs only), ultrasonication of the mixing water, representing a solution of NMCSs, was employed prior to mixing it with the dry cement powder. In the latter case (NMCSs + F127), the admixtures were just added within mortar casting in the relevant concentrations. Specimens in which only F127 was added to the mortar mixture were an additional, i.e., control case, for comparative purposes to the samples NMCSs + F127, where F127 was a dispersion agent. The designation and details on sample preparation for the mortar specimens are shown in Table 6.1.

The electrical resistivity of mortar was measured using an AC "2-pin method," where the "pins" were brass mesh pieces with dimensions equal to the sides of the mortar cubes [31]. An R-meter was used to record the electrical resistance of the mortar, measured by applying an alternating current of 1 mA at a frequency of 1 kHz. Electrical resistivity was calculated using Ohm's law: $\rho = R.A/l$, where ρ is the mortar resistivity (in Ohm.cm), R is the electrical resistance of mortar (in Ohm), A is the cross section of the mortar cube (in cm^2) and l is the length of mortar (in cm). The electrical resistivity was monitored immediately after casting and up to 36 days of curing. The tested specimens were maintained and covered in the mold during the whole test period (minimizing environmental effects) and were exposed to constant external conditions.

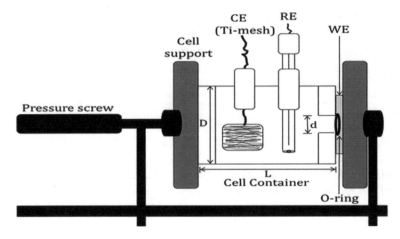

Fig. 6.2 Experimental setup of the three-electrode cell arrangement

Standard compressive strength tests were performed on the 40 × 40 × 40 mm mortar cubes at the hydration age of 36 days. Four replicate mortar specimens were tested.

6.2.3 Steel Electrodes, NMCSs, Model Medium and Sample Designation

Steel electrodes (low carbon St37) with a defined surface area of 0.3 cm^2 were ultrasonically degreased in ethanol, grinded with successive grades of SiC emery paper up to 4000 grade and then polished using 3 and 1 μm diamond paste to mirror-like surface. The electrodes were tested in model solutions of 0.9 M NaOH and 0.1 M NaOH of pH 13.9 and 12.8, respectively, at 25 °C. The electrochemical response of steel in the presence of 0.016% NMCSs was studied in these model solutions, using a flat three-electrode electrochemical cell. The schematic presentation of the employed cell is shown in Fig. 6.2.

Saturated calomel electrode (SCE) was used as a reference electrode, and MMO Ti- mesh as a counter electrode. For each exposure condition, two identical steel specimens (replicates) were tested. Similarly to the cement-based samples, the NMCSs were dispersed in two manners: through ultrasonication or through adding F127 (10 wt.%) as a dispersing agent. The samples designation, corresponding to each studied environment, is given in Table 6.2.

Prior to adding the NMCSs in the relevant solutions, they have been characterized in an identical medium by measuring their Zeta-potential, hydrodynamic radius and particle size distribution, using dynamic light scattering (DLS). The morphology of the NMCSs after carbonization was observed by a transmission electron microscope (JEM-2010).

Table 6.2 Test solution composition and sample designation

NaOH conc. M	pH	NMCS wt.%	F127 wt.%	Designation
0.1	12.8	0	0	A1
		0.016	0	A2
		0	10	A3
		0.016	10	A4
0.9	13.9	0	0	B1
		0.016	0	B2
		0	10	B3
		0.016	10	B4

6.2.4   Electrochemical Methods

Electrochemical measurements were employed to follow up the progressive changes in the electrochemical response of the steel electrodes in control conditions (no additives to the model medium) and in the presence of NMCSs and F127 during 3 days of exposure. The open circuit potential (E_{OCP}) was monitored at different intervals. Electrochemical impedance spectroscopy (EIS) was carried out using Metrohm Autolab-Potentiostat PGSTAT30. EIS measurements were recorded under potentiostatic control at OCP, from 10 kHz down to 10 mHz. An AC perturbation voltage signal of 10 mV (rms) amplitude was applied.

Periodic measurements of the polarization resistance (R_p) were carried out by potentiodynamic measurements from -20 mV to $+20$ mV versus OCP, at a scan rate of 10 mV/min. The corrosion current density (i_{corr}) was calculated by using the Stern-Geary eq. [32], $i_{corr} = B/R_p$, where B is the Stern-Geary constant. A value of $B = 26$ mV, for active carbon steel, was used and 52 mV for passive steel [33, 34].

At the end of the test, the specimens were subjected to cyclic voltammetry (CV). Ten CV cycles were performed from -1.2 V vs SCE to $+0.9$ V vs SCE, to cover all reduction and oxidation processes of interest, form hydrogen evolution to water oxidation and oxygen evolution. The scan rate was 1 mV/s.

6.3   Results and Discussion

6.3.1   NMCSs Characterization

The results from DLS and Zeta-potential measurements for the NMCSs in different model solutions (identical to those used for electrochemical tests later on) are listed in Table 6.3. In alkaline environment of pH 13.9 (0.9 M NaOH) and 12.8 (0.1 M NaOH), the average NMCSs particle size varied from 296 to 187 nm, respectively.

In alkaline solution with pH 12.8 and 13.9 (0.1 M and 0.9 M NaOH), the addition of 10% F127 was considered sufficient to ensure the formation of F127 micelles at

Table 6.3 Characterization of NMCSs in alkaline media

Test solution	pH	Zeta potential / mV[a]		DLS / nm
		1st	2nd	
0.9 M NaOH + NMCSs + F127	13.9	1.7	2.1	296
0.1 M NaOH + NMCSs + F127	12.8	1.3	0.98	187

[a]Two measurements were performed in two solutions of the same composition

room temperature, especially in solutions with pH 13.9 [35]. In addition to the possible adsorption of F127 on NMCSs surface, micelles formation was expected to induce additional dispersion of NMCSs in the bulk volume. In solutions of pH 12.8, a clear decrease in the F127 micellization and the adsorption of nonionic surfactant, F127, on NMCSs active surfaces, was reported, which in turn affected dispersion of NMCSs particles. In this case, NMCSs particles with small radius ($\approx$ 187 nm) were easily dispersed; however, bigger aggregates were settled down. In solutions of pH 13.9, the dispersion efficiency was expected to be higher and larger NMCSs particles were dispersed easily in solutions of 0.9 M NaOH. In addition to the expected higher micellization degree of F127 in 0.9 M NaOH solutions, the higher OH^- concentration could lead to more preferential adsorption of OH^- ions on NMCSs. Consequently, NMCSs dispersion and agglomeration, together with higher particle size variation, were expected compared to those in pH 12.8. As reported [36, 37], the critical micelle concentration (CMC) of F127 is strongly dependent on different parameters such as the pH, temperature and ionic strength. Solubility and CMC of F127 strongly decrease as the temperature increases [38, 39]. Additionally, with a pH increase, a reduction in the micelles radius was reported [39].

Pluronic F127 was used in this study as a dispersing agent only. Therefore, evaluating possible effect of F127 on steel electrochemical performance and/or the mortar bulk matrix was not an objective of this work. However, in view of the above considerations, related to behavior of F127 alone, the choice to evaluate it as a separate addition to model solutions, as well as an admixture to mortar, is logic and hence presented in this work for comparative purposes.

Some authors [40–42] argue that nitrogen-adjacent carbon atoms are positively charged and act as active sites for the adsorption of oxygen and/or negatively charged species as OH^-. In alkaline solutions with pH higher than 13, Wan et al. [42, 43] have concluded that the hydroxyl ion sorption on N-doped carbon surfaces proceeded a quick and reversible process. The positive Zeta-potential values of NMCSs are mainly related to positively charged N-atoms. The slight variation in the NMCSs Zeta-potential in solutions of pH 13.9 and 12.8 was probably related to preferential adsorption of cations or anions within the double layer of the active surface [38, 42, 43]. On the other hand, the pH variation induced changes in the nitrogen-oxygen surface active functional groups such as hydroquione-/quinone-like groups in NMCSs [42]. This type of variation of the nitrogen-doped sites on the carbon spheres (Pyridine or Pyrolic N- atoms) is pH-dependent, yielding a considerable change in the chemical state of the functional groups and active sites on the NMCSs, which causes variation in NMCSs Zeta-potential and average particle size, as listed in Table 6.3.

Considering reported and the hereby observed alterations in charge, size and overall behavior of the NMCSs in the tested model solutions, variations in the passive layer formation on the tested steel surface were expected. Additionally, positive charge of the NMCSs would account for certain limitations on the steel surface in their presence, e.g., positively charged particles can be expected to adsorb on anodic locations (negatively charged in a corrosion cell) and limit the oxidation reactions. Limitations in oxidation, on the other hand, although positive in view of global corrosion resistance development, will impede the passive layer formation and stabilization on the steel surface in the hereby tested environment. Therefore, changes in particles' charge are important in view of the global electrochemical state and will determine the performance of steel in the tested model environment, although steel in alkaline medium (not exceeding pH of 13.7) is expected to be in a passive state. Next to the above and in the case of (reinforced) cement-based materials, the process of cement hydration and distribution of hydration products are expected to be influenced by admixed NMCSs of varying charge. This would be especially the case if early stages of cement hydration are compared to later ages. All these phenomena are not subject to this work and need further in-depth investigation. However, the hereby reported results can be considered preliminary in view of the above aspects and provide an indication for the expected behavior and performance of both steel and cement-based bulk matrix.

6.3.2 Results from Preliminary Tests Bulk Matrix Properties

As aforementioned, reinforced cement-based material can be subjected to certain modification, e.g., by introducing admixtures to the bulk cementitious matrix, investigating the global material properties of the modified bulk material and the corrosion resistance of embedded steel is of equal importance. This is especially the case when an admixture aims to simultaneously affect the concrete microstructural properties (e.g., porosity, permeability and resistivity) and the corrosion resistance of embedded steel, with the final goal to achieve a reinforced concrete system of a superior performance. To that end, a sequence of studies is normally to be followed, before the optimum choice for tests in reinforced concrete is determined. For example, the admixtures are tested for their effect on cement-based bulk matrix only in plain (nonreinforced) specimens, whereas their effect on steel electrochemical response is tested in simulated aqueous environment [33, 43, 44]. In that manner the contribution of many effects in a system of high heterogeneity, as reinforced concrete, is separately evaluated. Therefore, with respect to this work, preliminary tests on global material properties of mortar specimens were performed, when admixtures as the hereby discussed NMCSs were involved.

Electrical resistivity measurements of $40 \times 40 \times 40$ mm cubic mortar specimens were employed as a rapid nondestructive testing method that allows investigating

Table 6.4 Electrical resistivity of mortar cubes at curing time intervals from 0.5 to 36 days

Curing time/ d	Electrical resistivity / Ω.cm			
	Blank	NMCS	NMCS + F127	F127
0.5	142	138	188	174
1	270	271	327	307
5	546	410	963	764
10	1159	878	2012	1618
20	1456	1119	2218	1889
36	1577	1394	2679	2152

Table 6.5 Compressive strength of mortar specimens with different additives after 36 days curing

Mortar specimens	Maximum strength / MPa	
	Mean value	Stand. Dev.
Blank	40.9	3.1
NMCS	54.3	3.9
NMCS + F127	11.5	2.5
F127	16.8	1.6

the quality of the mortar properties. The variations in the electrical resistivity of different mortar specimens with and without 0.025 wt.% NMCSs at different curing time and age respectively are listed in Table 6.4.

As expected, the electrical resistivity increases with cement hydration and age for all tested specimens. What can be noted is that the early properties of the cement-based bulk matrix containing NMCSs are possibly significantly altered. This is reflected by lower electrical resistivity of NMCSs-modified mortar cubes, especially pronounced from initial stages (0.5–5 days) and further until 20 days. At the end of the testing period, 36 days of age, the NMCSs-containing mortar cubes maintained the lowest electrical resistivity values among all tests groups – Table 6.4. In contrast, the highest resistivity values were recorded for mortar cubes where NMCSs were admixed together with F127. This can be attributed to (i) increased cement hydration and densification of the bulk matrix or (ii) decreased ionic mobility in the pore water, blocking of connected pore network paths and/or altered water balance in the presence of nonionic F127. These will end up in changing the process of cement hydration and could have a negative or positive effect, e.g., retardation or acceleration of cement hydration and pore network development, formation of insulated pore space (e.g., air-filled voids). Since the presence of only F127 results in slightly lower electrical resistivity values, compared to NMCSs + F127 (Table 6.4), F127 alone is most likely not responsible for the observed highest electrical resistivity in specimens, containing F127 and NMCSs together.

Additionally, if electrical resistivity values are correlated to compressive strength development (Table 6.5), the influence of admixtures can be judged in terms of global material properties development.

As can be observed in Table 6.5, significantly lower compressive strength was recorded for the mortar matrix, containing F127 alone or F127 together with NMCSs, i.e., in the range of 12–17 MPa compared to 41 MPa for the control case. This result is in contradiction with the higher values of recorded electrical resistivity (Table 6.4) for those two cases, considering the general perception that higher electrical resistivity accounts for a denser matrix, whereas a denser or more matured cement-based material would have a higher compressive strength. Next to that, the specimens containing NMCSs only exhibit the highest compressive strength – 54 MPa. However, these specimens show the lowest electrical resistivity values. In the case of NMCSs alone, the opposite trends of electrical resistivity and compressive strength development can be explained with increased conductivity of the matrix due to the presence of NMCSs, being a carbon-based material of high conductivity as an intrinsic property. The highest compressive strength, though, would be rather related to phenomena linked to cement hydration and formation and/or (re) distribution of calcium-silicate hydrates.

Apparently, competing mechanisms of altered cement hydration, nucleation effects and/or redistribution of hydration products, connected and disconnected pore network, play a role in the development of the cement-based microstructure when both NMCSs and F127 are present. These aspects need further in-depth investigation, which was not subject to the present work. For the purposes of this study, the recorded global material properties are sufficient to justify the selection of the investigated admixtures and their concentration respectively, in order to account for corrosion tests in reinforced mortar and the objective of simultaneous steel corrosion control and improved bulk matrix characteristics. To that end, and at this stage, it can be concluded that the NMCSs would possibly result in superior properties of a reinforced concrete system. However, F127 might not be the best and suitable option as a dispersing agent for the carbon spheres, if the hereby discussed results for compressive strength and electrical resistivity development are considered.

6.3.3   Electrochemical Performance of Steel in the Presence of NMCSs

6.3.3.1   Open Circuit Potential (OCP) and Corrosion Current Density (i_{corr})

The open circuit potential (OCP) and the corrosion current density (i_{corr}) of carbon steel have been monitored in the course of 3 days of continuous exposure to 0.1 M (pH 12.8) and 0.9 M (pH 13.9) NaOH solutions; the obtained data are displayed in Fig. 6.3.

In 0.1 M NaOH solutions (samples A1–A4), more anodic OCP values were measured and i_{corr} stabilized at low current densities after 72 h of immersion in this alkaline medium (Fig. 6.3a, b). This variation in OCP and i_{corr} is related to the formation of electrochemically stable passive film. Similar mechanisms would hold for

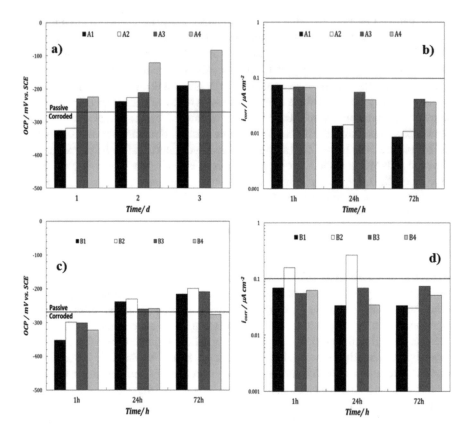

Fig. 6.3 Variation of the open circuit potential (OCP) and corrosion current density (i_{corr}) of the tested steels immersed in 0.1 M NaOH (solutions A) and 0.9 M NaOH (solutions B) after 1, 24 and 72 h. **a**, **c**) E_{corr}, and **b**, **d**) i_{corr}

the steel tested in 0.9 M NaOH environment (solutions B), although more cathodic OCP values were recorded for these specimens, accompanied by slightly higher corrosion current densities (Fig. 6.3c, d). This was as expected due to the aggressiveness of solutions B towards steel, i.e., pH 13.9 is a medium where carbon steel would present active state until passive layer formation will overcome active oxidation. For specimens A1 and A2, the recorded anodic shift in OCP and low i_{corr} values with treatment in 0.1 M NaOH solutions followed the generally expected trend of steel passivation in alkaline medium of pH 12.8. This is due to the gradual development of an oxide film on the surface, as also reported [45–47]. An OCP shift to more anodic values generally reflects a passive film acquiring a stable thickness over time [48]. The addition of 0.016% NMCSs without F127 (solution A2) did not significantly affect the passive film development on the steel surface, i.e., specimens A2 showed similar response to the steel in solutions without additives (A1), with only slight variations in OPC and i_{corr} values. The more noble potential of A2, compared to A1, Fig. 6.3a, would account for a generally improved corrosion resistance.

However, the i_{corr} values as derived for A2 were slightly higher than those for A1 (Fig. 6.3b). The variation between A1 and A2 solutions could be assigned to the effect of NMCSs on the oxygen reduction reaction, resulting in competing mechanisms of passive layer growth on the one hand and catalyzed reduction, i.e., enhanced overall oxidation rate, on the other hand. While the former will account for the establishment of a stable passive state, the latter will lead to increase of the overall corrosion current density. This mechanism is independent from the possible adherence of NMCSs (positive charge) to anodic locations (negatively charged) and effect on the anodic reaction itself. Overall, enhanced reduction, together with blocked anodic areas would result in higher corrosion activity, since more time will be needed for the development of an electrochemically stable and uniform passive film. Therefore, the corrosion activity of A2 ends up slightly higher than that for A1, although a clear trend towards an electrochemically stable passive state was observed, based on OCP and i_{corr} evolution (Fig. 6.3a, b).

The response of specimens A3 and A4 showed a clearly different trend. In both cases, although OCP exhibited ennoblement with time of treatment, especially for the case of A4 (Fig. 6.3a), higher corrosion current was recorded, compared to A1 and A2 (Fig. 6.3b). Obviously, the presence of F127 in both A3 and A4 cases was responsible for the observed behavior. The presence of F127 only, specimen A3, hindered the passive film formation. This is reflected by almost constant OCP values for A3 throughout the test, i.e., substantial ennoblement was not recorded as would otherwise reflect improvement of passivity. The trend of this OCP evolution for A3 was in line with the nonsignificant trend of current density decrease (Fig. 6.3b), i_{corr} values remaining almost one order higher than those for A1 and A2 at the end of the test. The effect of F127 can be also observed for the case of A4 specimens, where despite the very noble OCP values (Fig. 6.3a), the current densities were, similarly to A3, higher than those for A1 and A2 cases. In case of A4 specimens, competitive mechanisms were involved, where the positive effect of NMCSs was counterbalanced by F127. In other words, enhanced oxygen reduction and initially blocked active areas on the steel surface (NMCSs effect), otherwise resulting in a faster development of a stable layer, were counteracted by the presence of F127 (inducing barrier effect); F127 blocked the steel surface, resulting in noble potentials but a nonstable electrochemical state and impeded formation of a passive film. Therefore, in the case of specimens A4 the higher anodic shift in OCP and high i_{corr} values, compared to A1 and A2 solutions, were denoted to a possible formation of an adsorbed layer of nonionic F127 triblock copolymer containing NMCSs on the steel surface. This layer was actually as observed by ESEM examination, as will be presented further below.

The OCP values for steel treated in solution of 0.9 M NaOH (pH of 13.9), (Fig. 6.3c, d), were more cathodic than those in solution of 0.1 M NaOH (pH of 12.8). This was as expected due to the initially more active steel surface at the higher pH, as thermodynamically justified.

For specimens B1 and B2 similar OCP values were measured, while corrosion current densities were higher in the presence of NMCSs, i.e., for B2 (Fig. 6.3c, d). This was related to the effect of NMCSs, inducing high catalytic activity for oxygen reduction and subsequently enhanced oxidation rate. Once the film developed on the steel surface, the corrosion activity in B2 was reduced, ending up with even lower i_{corr} values, compared to B1 (Fig. 6.3d). The increased corrosion resistance of B2 was reflected by more noble potentials at the end of the test, compared to all other cases (Fig. 6.3c).

Similarly to the A samples, blocking the steel surface was relevant for the F127-containing B samples as well, specimens B3 and B4. Both OCP and i_{corr} records account for reduced corrosion resistance in B3 and B4 samples. The responsible mechanisms here were analogical to those described for specimens A, however, even more pronounced due to the higher corrosivity of the medium, i.e., pH 13.9.

The effect of pH only on the steel electrochemical response, and passive film formation, respectively, can be clearly observed by the comparison of OCP and i_{corr} values of the treated steel in 0.1 M NaOH (A1) and 0.9 M NaOH (B1) solution. In alkaline solutions of pH 13.9 (B1), higher i_{corr} values were registered. It is well known [47–50] that the reduction of Fe^{3+} to Fe^{2+} generally constitutes an additional cathodic reaction, which would affect both OCP and i_{corr} values. The contribution of Fe^{2+} species was expected to be more evident when the potential of the passive film shifted to more negative values [50]. Consequently, the high i_{corr} values in 0.9 M NaOH solutions can be mainly related to the accumulation of Fe^{2+} species within the passive film. However, when the film is exposed to solutions of pH 12.8, which is the case of 0.1 M NaOH solution (A1) in this work, more anodic OCP and lower i_{corr} values would be mainly related to enrichment of the passive film in Fe^{3+} oxides/hydroxides [49, 50].

In the presence of 0.016% NMCSs + 10% F127 (B4), the presence of positively charged NMCSs would limit the Fe^{2+} to Fe^{3+} oxidation, which in turn will lead to increasing the Fe^{2+}-oxide species. This explains the more cathodic potential measured in case of B4 solutions. This cathodic shift in OCP was not observed in solutions B3 with only F127.

6.3.3.2 Electrochemical Impedance Spectroscopy (EIS): Qualification of Response

Electrochemical impedance spectroscopy (EIS) was performed for the tested steel in 0.1 M NaOH (designation A) and 0.9 M NaOH (designation B) solutions and samples respectively. The EIS response in a Nyquist plot format for all specimens, treated in the relevant solutions with and without additives, is presented in Figs. 6.4 and 6.5. In 0.1 M and 0.9 M NaOH solutions without additives (A1 and B1), the limit of impedance values at the low frequency range increases with immersion time, reflecting the spontaneous passivation of steel in alkaline media, Figs. 6.4a and 6.5a. The response in 0.9 M NaOH, Fig. 6.5a, depicts lower corrosion resistance (lower global impedance |Z| values) in accordance with the expected effect of solutions of pH > 13.7, i.e., steel in more active initial state for cases B, compared to medium of pH 12.8, cases A, where only passivity stabilization would be relevant. These observations are in line with the recorded evolution of OCP and i_{corr} for

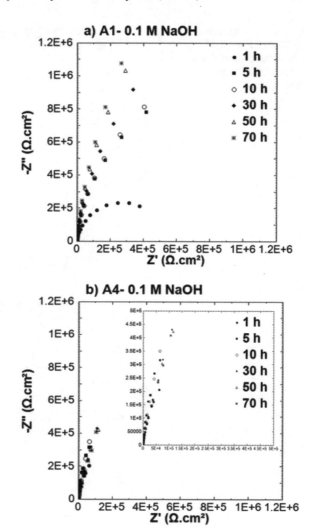

Fig. 6.4 EIS response of St37 after 1, 5, 10, 30, 50 and 70 h of immersion at pH 12.8 in solutions A1 and A4

specimens in solutions A1 and B1 with time of treatment, Fig. 6.3a, c. The EIS results also confirm the previously discussed hypothesis for a more electrochemically stable passive film in conditions of pH 12.8, resulting from predominance of Fe^{3+} species. Furthermore, the higher stability of the passive layer on the steel surface in A1 conditions is evident from the already more pronounced increase of impedance magnitude towards the end of the test (50–70 h) if compared to the EIS response for steel in solution B1 – compare responses for A1 and B1 in Figs. 6.4a and 6.5a.

If the EIS response of steel treated in solutions containing 0.016 wt.% NMCSs +10% F127 (solutions A4 and B4) is compared to additives-free solutions (A1 and B1), the main observations are as follows: the overall impedance in case of solutions A4 was lower than that for additives-free solutions in the same medium (A1) –

Fig. 6.5 EIS response of
St37 after 1, 5, 10, 30, 50
and 70 h of immersion at
pH 13.9 in solutions B1
and B4

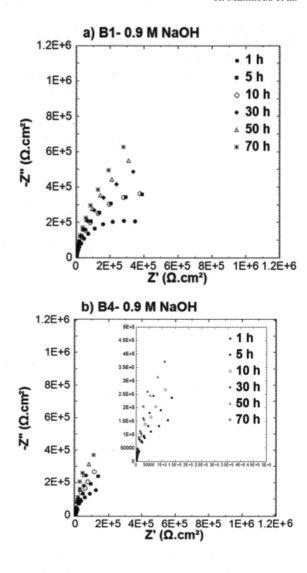

Fig. 6.4a, b. Furthermore, no increase in the overall impedance with treatment in
solution A4 was observed, where the limit of global impedance at the low frequency
range remained almost constant after 10 h and until the end of the test (Fig. 6.4b).
The observed response accounts for possible limitations within the passive film formation, development and stabilization in the case of A4, which is obviously related
to the presence of both NMCSs + F127. The almost constant EIS response for A4
throughout the test reflects a state of counterbalanced diffusion limitation (barrier
effect of F127) and passive layer formation. This is in line with the previously discussed simultaneous action of positive NMCSs effect and negative F127 effect,
which resulted in OCP ennoblement but a delay in a stable passive film formation.

In the case of the B4 samples, the originally higher steel activity in 0.9 M NaOH solutions impeded the barrier effect of F127 and altered the action of NMCSs. The result was a nonstable response over time, with a gradual increase in global magnitude of impedance. The corrosion resistance of B4, however, remained lower than that for the B1 specimens (Fig. 6.5a, b). Here again, the addition of NMCSs + F127 limited the development of a stable passive film, although the relevant mechanism was entirely different from that in solutions A4. In both cases where particles were present – A4 and B4, the overall impedance ends up lower compared to particles free solutions – A1 and B1. This is well in line with the recorded corrosion current densities towards the end of the test (Fig. 6.3b, d, where the corrosion current densities for steel in solutions A4 and B4 were higher than these for steel, treated in solutions A1 and B1.

Figure 6.6a, b depicts the Nyquist diagrams of the impedance spectra for all tested steel specimens at the end of the test. In 0.1 M NaOH (pH 12.8) – solutions A1 and A2, the addition of 0.016% NMCSs showed nonsignificant effect on the impedance response (Fig. 6.6a). For steel treated in the presence of F127 only, specimens A3, a less capacitive behavior compared to solutions A1 and A2 was observed. This decrease in the overall impedance of the tested steels in solutions A3 is related to retardation in passive film formation, induced by diffusion limitations at the steel/solution interface in the presence of F127. A mixed charge transfer/mass transfer mechanism of the electrochemical reaction in A3 is evident by the depression (distortion) of the semicircular response of A3 in the low frequency domain (Fig. 6.6a, showing an initial capacitive-like response, followed by inclination of the curve to the real axis). On the other hand, in case of steel treated in solutions A4 with NMCSs +10% F127, relatively higher impedance than in case of A3 solutions was measured, Fig. 6.6a. This is reflected by the more pronounced capacitive-like behavior of A4 (response inclined to the imaginary axis) which is also in agreement with the more noble OCP values, measured for A4, compared to A3 at the same exposure conditions, Fig. 6.3a. Since for A4 the competition of passive layer formation (at pH 12.8 and NMCSs effect) counteracted the limitations to stable passivity (F127 affect), the global magnitude of |Z| ends up higher than that for A3, but lower than A1 and A2. The result is well in-line with the recorded corrosion current density from LPR method (Fig. 6.3b).

Regarding 0.9 M NaOH solutions (pH 13.9), the tested B specimens (B1–B4) showed a relatively lower overall impedance, when compared to A-specimens. The comparison of the steel EIS response for B1 and B2 cases indicated that, the addition of 0.016% NMCSs (B2 solution) induced a decrease in impedance compared to the B1-solution, Fig. 6.6b. Concerning EIS response of the immersed steel in B4 solution, more resistive behavior than in B1, B2 solutions, was observed, while low impedance values were derived in the case of F127 only, i.e., specimens B3. This trend for B3 is similar to that obtained in A3 solution, Fig. 6.6a, b. The effect of NMCSs and F127 together (B4) was again seen in the increase of corrosion resistance, compared to F127 alone (B3). In contrast to A4 specimens, the effect of NMCSs + F127 together, as in B4 specimens in medium of pH 13.9, was more pronounced and not entirely in line with the results in Fig. 6.3b, c, where cathodic

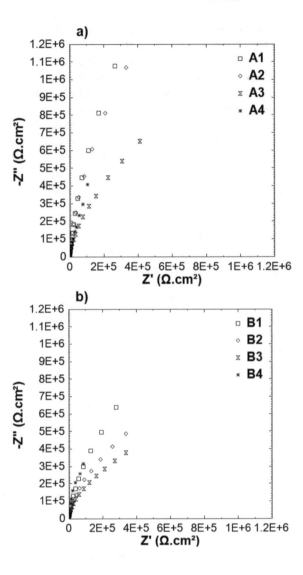

Fig. 6.6 Nyquist plots for
St37 after 70 h of
immersion in all tested
solutions

values for B4 were recorded. However, as previously discussed, the OCP for B4
remained almost constant at the later stages of the test. This was discussed to be
related to a hold-back of stable passive film formation due to limitations at the steel/
solution interface in the presence of F127, on one hand, and enhanced redox activity
due to the presence of NMCSs and the higher pH of the B solution, on the other
hand. Hence, although similar response was monitored in case of A4 solutions, the
nature of the mechanisms involved in A4 and B4 are different, as also reflected by
both electrochemical response and surface properties (presented further below). The
reason behind these observations is related to the effect of pH on F127 itself,
together with the effect of pH on the dispersion of the NMCSs. These transforma-

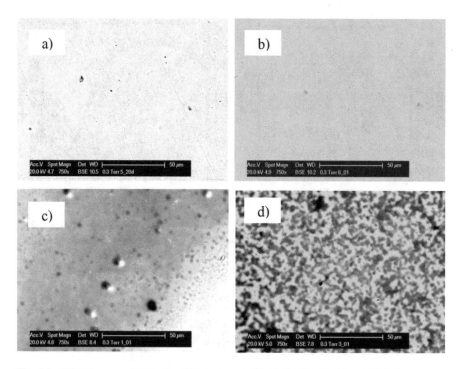

Fig. 6.7 ESEM images of St37 after 70 h treatment: (**a**) A1, (**b**) B1, (**c**) A4 and (**d**) B4

tions, although potentially of high importance in view their effect on electrochemical state, were not subject to this work. What can be clearly stated, though, is that at pH of 12.8, the NMCSs particles in the bulk of the solution, and presumably adjacent to the steel surface, were of a smaller size (approx. 200 nm), while the dispersed particles in pH 13.9 were larger – approx. 300 nm (Table 6.3).

The particle size distribution is perhaps a large contributing factor to the observed electrochemical response. Smaller particles exhibit larger surface area. This, together with the previously introduced electrocatalytic, etc. properties of NMCSs, will logically and irrespective of all other factors determine the oxidation/reduction reactions on the steel surface. These reactions would be facilitated in specimens A and would result in a faster stabilization of passivity, if compared to specimens B – which was as observed. Clearly, a more in-depth investigation on the interactions of NMCSs, with and without F127, with the steel surface is needed in order to confirm or reject the above hypothesis.

The above discussed phenomena with regard to passive film formation, or limitations thereof, together with the recorded EIS response, i_{corr} and OCP, are well in line and supported by microscopic examination of the steel surface at the end of the test. The visual observation (via electron microscopy) reveals the relatively clean surface for steel treated in additives-free solutions – A1 and B1, Fig. 6.7a, b. In contrast, the appearance of steel treated in A4 and B4 solutions varies significantly, Fig. 6.7c, d. In the former case (A4 solutions), the steel surface appears to be smooth and in

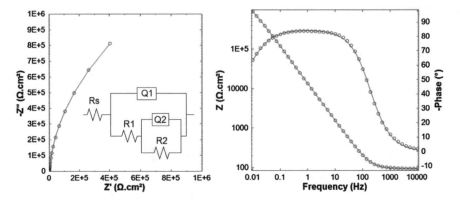

Fig. 6.8 EIS response obtained for carbon steel after 10 h of immersion in A1-solution (0.1 M NaOH). Experimental data (*dots*) and fitting (*solid line*), using the equivalent circuit $R_e(R_1Q_1(R_2Q_2))$, where $R_e = 96\Omega$, $R_1 = 621$ kΩ cm^2, $Q_1 = 5.1 \times 10^{-5}$, $n = 0.94$, $R_2 = 211$ kΩ cm^2, $Q_2 = 1.7 \times 10^{-5}$, $n = 0.9$, and $\chi^2 = 0.005$

similar to A1 condition with the difference of randomly distributed particles, obviously adhering to the substrate. In the latter case (B4), the steel surface appears to be covered by a dense composite layer, containing both NMCSs and F127.

6.3.3.3 EIS Response: Brief Intro to Quantification

The recorded EIS response for all investigated cases was discussed in the previous sections in view of qualification of electrochemical state. Qualitative interpretation of EIS data is possible and well accepted as an approach, providing a rapid evaluation of the overall corrosion resistance for the tested systems. The quantitative evaluation of EIS is relatively more complex. This is in view of the necessity to account for exact physical meaning and discuss in detail the relevant electrical parameters used within fitting of the experimental data. This type of quantification is not subject to this work. An introduction to the quantification of the recorded EIS response is provided only, together with the derived polarization resistance (R_p) values from EIS as an AC method. The objective was to provide a comparison with R_p values derived from DC electrochemical tests in view of evaluating the global corrosion state of the tested specimens.

Figure 6.8 presents the EIS response, fit and equivalent electrical circuit employed for a steel specimen treated in A1-solution after 10 h of conditioning. Figure 6.8 is representative for all investigated cases. The best-fit parameters were derived by using an equivalent electrical circuit of two hierarchically arranged time constants in series with the electrolyte resistance, i.e., $R_e(R_1Q_1(R_2Q_2))$, where Re is the solution resistance; the high frequency time constant ($R_1 \cdot Q_1$) is associated with the charge transfer resistance and pseudo-double layer capacitance; the low frequency time constant ($R_2 \cdot Q_2$) is related to the redox transformations, mainly Fe^{2+}/Fe^{3+}, in the surface layer [44, 51, 52]. The replacement of pure capacitance with a constant phase element (Q) is largely accepted in literature [44, 51] because of inhomogeneities at different levels, i.e., steel surface roughness, product layer heterogeneity, etc. The global polarization resistance (R_p) values can be derived from EIS via simplification, using the relation of $R_p = R_1 + R_2$ [44]. A good agreement between the model and the experimental data was found, reflected by acceptable error per element and for the overall fit of the full response (Fig. 6.8).

The comparison of the global polarization resistance, R_p, measured from DC measurement (LPR) and the one, derived from EIS is shown in Fig. 6.9a, b. The R_p values measured by EIS data were higher than those recorded via LPR. Experimentally, the polarization resistance measurements were done at 4 mHz (±20 mV around OCP at 10 mV/min), below the lowest frequency limit considered for EIS measurements (10 mHz) [53, 54]. Thus, the measured R_p form DC measurements are affected by both R_1 and R_2 parameters, which are related to the charge transfer resistance and redox processes, which occurred in the passive film layer, respectively. The deviations in the R_p measurements by EIS and DC method are more relevant after 72 h immersion in both 0.1 and 0.9 M NaOH, A and B solutions. This is related to the employment of different electrochemical AC and DC techniques in case of EIS and LPR, respectively.

It should be mentioned that both the R_p values measured by EIS or DC techniques were reproducible and similar variation trends in parallel exposure conditions was observed Fig. 6.9a, b.

In additive free solutions, A1 and B1, the polarization resistance of the tested steels in these solutions was improved by aging, as shown in Fig. 6.9a, b. Furthermore, in agreement with the aforementioned results (Figs. 6.3 and 6.5), the measured resistance in B1 solutions was relatively lower than in case of A1 solutions. This was related to the pH effect on the grown passive film of the treated steels in additive-free A1 and B1 solutions.

The tested steels in solutions A2 and B2, in the presence of NMCSs, showed similar R_p values to those measured in case of solutions A1 and B1, respectively, Fig. 6.9a, b. In 0.1 M NaOH solutions with 10% F127 (solutions A3), the R_p values showed nonsignificant variation after 24 h of immersion. These results confirm that the passive film formation was hindered in the presence of F127 at the steel/solution interface. On the other hand, the presence of NMCSs with F127 (solution A4) resulted in a slight increase in R_p values, observed after 72 h, Fig. 6.9a. Concerning the tested steels in B4 solutions, the R_p values shows similar variation trend to B1 and B2 solutions, reaching almost the same R_p values after 72 h aging, Fig. 6.9b. These results were not as the observed results in solutions A4, where the R_p showed no significant variation after 24 h, Fig. 6.9a. This can be explained in regards to the variation in dispersed NMCSs properties in B4 solutions, where the NMCSs was expected to be preferentially adsorbed on the anodic zones due to the variation in the NMCSs charge density (Table 6.3). In B4 solutions with pH 13.9, the anodic zones are expected to be more dense than in the case of A4 solutions (pH 12.8). This could explain the visual changes of the adsorbed layer on the steel surface immersed in A4 and B4 detected by ESEM observation, shown in Fig. 6.7c, d.

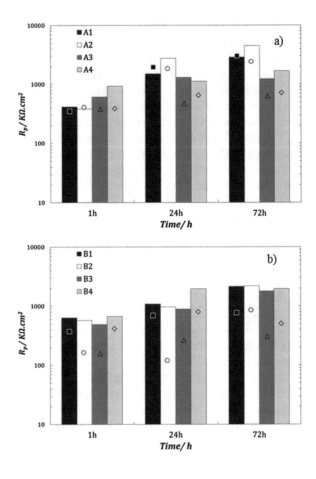

Fig. 6.9 Comparison of the polarization resistance, R_p, measured by DC and AC methods (LPR and EIS): (**a**) 0.1 M NaOH solutions (A1–A4) and (**b**) 0.9 M NaOH solutions (B1–B4) after 1, 24 and 72 h. *Column*: R_p from EIS; *Symbols*: R_p from LPR

6.3.3.4   Cyclic Voltammetry (CV)

Carbon steel behavior in 0.1 and 0.9 M NaOH solutions of different pH, with and without NMCSs, was analyzed by performing CV after 3 days of immersion.

Figure 6.10a, b present the voltammograms corresponding to the last cycle of the tested carbon steel in 0.1 M NaOH solution (A1 and A4) and 0.9 M NaOH solutions (B1 and B4), respectively. Three potential domains can be clearly distinguished: iron redox process domain, passivation region and oxygen evolution, Fig. 6.10a, b.

In the forward scan, the main iron activity peak appeared E ≈ −0.75 V vs SCE (peak I) was assigned mainly to magnetite formation [46, 55, 56]. Then, the current was maintained almost constant along the passivity domain (from −0.4 to +0.5 V) until reaching the oxygen evolution potential, E ≈ 0.55 V vs SCE (peak II). Fliz et al. [56] have indicated that at potential above −0.4 V vs SCE, the potentiodynamically grown passive film did not contain ferrous compounds and it mainly contained $Fe(OH)_3$ and/or FeOOH.

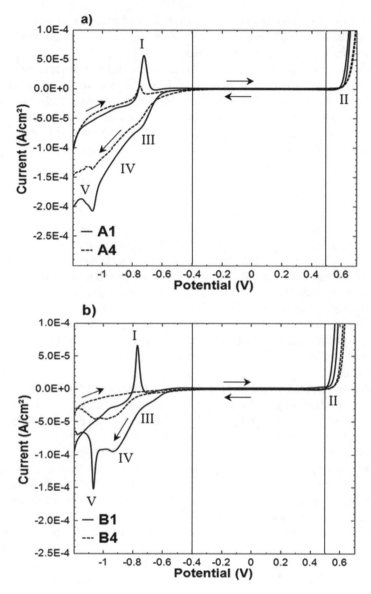

Fig. 6.10 Comparison of the 10th cycles of carbon steels after 3 days of immersion in (**a**) 0.1 M NaOH solution (A1–A4) and (**b**) 0.9 M NaOH (B1–B4)

In the cathodic scan, the formed iron oxide species reduced at three different cathodic potentials assigned as peaks III, V and IV, which corresponded to the reduction of Fe^{3+} species, Magnetite and Fe^{2+} species, respectively [46].

By comparing the cyclic voltammograms in solutions A1 and B1, a cathodic shift in peak I in solution B1 of pH 13.9 was monitored with respect to the same peak appearing in case of solution A1. Furthermore, the cathodic current densities associated to peaks III, IV and V were relatively higher in case of A1 solution when compared to solution B1, as shown in Fig. 6.3.

If solutions with additives (A4 and B4) are compared to solutions without additives (A1 and B1), a very well pronounced difference was observed for both 12.8 and 13.9 solutions. A clear decrease of the current densities associated to anodic and cathodic peaks was monitored. In addition, the anodic and cathodic peaks assigned to iron activity peaks were shifted to more cathodic potentials when compared to additive-free solutions (A1 and B1, respectively). The oxygen evolution peaks in case of solutions A4 and B4 were slightly shifted to more anodic potentials, reflecting the variation of steel/adjacent solutions interface properties in the presence of additives, compared to additives-free cases, Fig. 6.10.

In A4 solutions, the addition of 10% F127 + 0.016% NMCSs induced a clear decrease in peak I current with shift to more cathodic potential. Moreover, the cathodic current densities of the appeared peaks were lower than in case of A1 solution. In agreement with R_p data shown in Fig. 6.9a, this fact can explain the reason of nonsignificant variation in R_p in case of steel immersed in A4 solution. Mainly, because of the expected formation of less oxidized passive film with potentially lower magnetite content in the presence of 10% F127 + 0.016% NMCSs. In addition, a limited reduction process was observed in case of A4 solution, which confirms the possible blocking of the cathodic sites. This blocking did not induce additional barrier properties because of its effect on the passive film chemical composition.

In case of the treated steels in B4 solutions, the presence of 10% F127 + 0.016% NMCSs caused a drastically lower anodic and cathodic activity peaks. This decrease in the iron activity peaks resulted from the formation of an adsorbed layer on the surface of the treated steel in B4 solutions, confirmed by surface examination using ESEM (Fig. 6.7d). In this case, the formed layer on the steel treated in solutions B4 induced additional barrier properties, which in turn explain the relatively high R_p values in the case of B4 solutions, especially after 1 and 24 h, Fig. 6.9b. However, after 3 days, similar R_p values to the control samples (B1) were monitored, which was higher than in case of A4-solutions, Fig. 6.9.

Fundamentally, the passive film has been described to have a bilayer structure, which consists of inner layer of Fe_3O_4 and an outer layer of γ- Fe_2O_3 [47, 52–57]. According to the point defect model (PDM), the passive film is a highly doped but graded defect semiconductor structure, in which the vacancies are assumed to act as the electronic dopants. The donor vacancy in the oxide films were Fe^{2+} interstitials and/or oxygen vacancies, and the former is the major donor since the rate of consumption is higher than the its generation [58–59]. The acceptor vacancies were the cation vacancies on cation sub-lattice in the barrier layer [58–59]. Based on the CV data, the mechanism of the potentiodynamically formed passive film could be explained as follows: at low potential (more cathodic than −0.7 V vs SCE), Fe_3O_4 was the main constituent of the passive film with relatively small amount of γ- Fe_2O_3 [57]. As the potential increased above −0.5 V vs SCE, more and more Fe_3O_4 was anodically converted to γ- Fe_2O_3 and γ- FeOOH [57].

The presence of NMCSs in the oxide/solution interface could play an important role in regulation of the electronic properties of the passive layer and its chemical composition. The presence of the carbon-based material in the passive film interface showed a tendency to accept the excess electrons from oxygen vacancies [60]. Therefore, the passive film formed in the presence of NMCSs was expected to have higher Fe^{2+} interstitials and/or oxygen, which is in agreement with the cathodic shift in peak I in solutions A4. Consequently, less γ- Fe_2O_3 and γ- FeOOH content would be related to A4, compared to the control cases (A1 solutions). The higher the carrier density, the greater is the passive film conductivity [60, 61]. Consequently, this increase in the conductivity led to a decreasing value of passive film R_p, compared to the additive-free case, which was in accordance with the measurements shown in Fig. 6.9. Furthermore, this can explain the decrease in the current densities associated to the anodic and the cathodic peaks, as appearing in the cyclic voltammograms in the presence of NMCSs, Fig. 6.10.

The treated steel samples in solutions B4 showed more pronounced blocking of the steel surface, where higher anodic current from the iron activity domain was measured. No clear oxidation peaks were observed. In the cathodic scan, the limitations of the NMCSs highly affected the iron oxide species reduction processes, because of the catalytic effect of NMCSs on oxygen reduction at the expense of iron reduction processes.

6.4   Conclusions

The electrochemical response of carbon steel in 0.1 and 0.9 M NaOH was controlled by the induced variation in the NMCSs properties, together with the pH value of the surrounding media. In alkaline solutions with pH 12.8, the presence of NMCSs at the steel/solution interface induced the formation of passive film with comparable to control conditions impedance magnitude. The fact was mainly related to the competition of passive layer formation and destabilization induced by the blockage of active areas on the steel surface by NMCSs.

In 0.9 M NaOH solutions (pH 13.9), the tested steels in B-solutions (B1–B4) showed a relatively lower overall impedance, when compared to A-solutions with pH 12.8. In the presence of NMCSs particles, a competing mechanism of active state, i.e., enhanced oxidation on the one hand, and particles adsorption on anodic sites and oxidation limitation, on the other hand, was relevant inducing larger fluctuations in impedance response and stabilization only towards the end of the testing period.

The use of F127 as a dispersion agent is not a suitable option for NMCSs in alkaline media due to its effect in limiting the passive film growth by the formation of an adsorbed layer on the steel surface.

Mortar bulk matrix properties were highly affected by NMCSs and F127. Lowest electrical resistivity values were monitored in mortar specimens mixed-in with only 0.025 wt.% NMCSs, which would result in better conduction of electrical current flow. However, the presence of F127 induces an increase in the electrical resistivity.

The addition of 0.025 wt.% NMCSs (with respect dry cement weight) improves the compressive strength. Nevertheless, the compressive strength of mortar specimens with F127 or NMCSs + F127 was three times lower than the control specimens. This decrease in compressive strength in the presence of F127 can be explained as the natural consequence of a progressive weakening of the matrix, probably caused by increasing porosity and/or hindering the cement hydration in the presence of nonionic F127.

References

1. Bertolini L, Elsener B, Pedeferri P, Polder R (2004) Corrosion of steel in concrete: prevention, diagnosis, repair, 1st edn. Wiley, Weinheim, pp 71–121
2. American Concrete Institute (1985) Committee 222, corrosion of metals in concrete. ACI J 81:3–32
3. Broomfield JP (2007) Corrosion of steel in concrete: understanding, investigation and repair. Spons Architecture Price Book, London
4. Balaguru PN, Shah SP (1992) Fiber reinforced cement composite. McGraw-Hill Inc., New York
5. Kowald T (2004) Influence of surface modified carbon nanotubes on ultra-high performance concrete. In: Schmidt M, Fehling E (eds) Proceedings international symposium on ultra-performance concrete. Kassel University Press GmbH, pp 195–202
6. Angst U, Elsener B, Larsen KC, Vennesland Ø (2009) Critical chloride content in reinforced concrete – a review. Cem Concr Res 39:1122–1138
7. Chernin L, Val DV (2011) Prediction of corrosion-induced cracking in reinforced concrete structures. Constr Build Mater 25:1854–1869
8. Lui Y, Weyers RE (1998) Modelling the time to corrosion cracking in chloride contaminated reinforced concrete structures. ACI Mater J 95:675–681
9. Koleva DA, de Wit JHW, van Breugel K, Lodhi ZF, van Wesring E (2007) Investigation of corrosion and cathodic protection in reinforced concrete. J Electrochem Soc 154:52–61
10. Mahmoud H, Alonso MC, Sanchez M (2012) Service life extension of concrete structures by increasing the chloride threshold using stainless steel reinforcements. In: Alexander MG et al (eds) Concrete repair, rehabilitation and retrofitting III. Taylor & Francis Group, London, pp 497–503
11. Pan Z, He L, Qiu L, Korayem AH, Li G, Zhu JW, Collins F, Li D, Duan WH, Wang MC (2015) Mechanical properties and microstructure of a graphene oxide-cement composite. Cem Concr Compos 58:140–147
12. Collins F, Lambert J, Duan WH (2012) The influence of admixtures on the dispersion, workability, and strength of carbon nanotuble-OPC past mixtures. Cem Concr Compos 34:201–207
13. Jo BW, Kim CH, Tea GH, Park JB (2007) Characteristics of cement mortar with nano-SiO2 particles. Constr Build Mater 21:1351–1355
14. Refait P, Jeannin M, Sabot R, Antony H, Pineau S (2013) Electrochemical formation and transformation of corrosion products on carbon steel under cathodic protection in seawater. Corros Sci 71:32–36
15. Carmona J, Garcés P, Climent MA (2015) Efficiency of a conductive cement-based anodic system for the application of cathodic protection, cathodic prevention and electrochemical

chloride extraction to control corrosion in reinforced concrete structures. Corros Sci 96:102–111
16. Duwell EJ, Todd JW, Butzke HC (1964) The mechanism of corrosion inhibition of steel by ethynylcyclohexanol in acid solution. Corros Sci 4:435–441
17. Yang W, Li Q, Xiao Q, Liang J (2015) Improvement of corrosion protective performance of organic coating on low carbon steel by PEO pretreatment. Prog Org Coat 89:260–266
18. EN 1990 (2002) Eurocode- basis of structural design. European Committee for Standardization, Brussels
19. Sanchez F, Sobolev K (2010) Nanotechnology in concrete- a review. Const Build Mater 24:2060–2071
20. Tang J, Liu J, Li C, Li Y, Tade MO, Dai S, Yamauchi Y (2014) Synthesis of nitrogen-doped mesoporpous carbon spheres with extra-large pores through assembly of diblock copolymer micelles. Angew Chem Int Edn 53:588–593
21. Yang T, Liu J, Zhou R, Chen Z, Xu H, Qiao SH, Monterio M (2014) N-doped mesoporous carbon shperes as the oxygen reduction catalysts. J Mater Chem A 2:18139–18146
22. Zhang LL, Gu Y, Zhao XS (2013) Advanced porous carbon electrodes for electrochemical capacitor. J Mater Chem A 1:9395–9408
23. Parveen S, Rana S, Fangueiro R (2013) A review on nanomaterial dispersion, microstructure and mechanical properties of carbon nanotube and nanofiber reinforced cementitious composites. Nanomaterials 2013:1–19
24. Yazdenbakhsh A, Grasley Z (2012) The theoretical maximum achievable dispersion of nanoinclusions in cement paste. Cem Concr Res 42:798–804
25. Konsta-Gdoutos MS, Metaxa ZS, Shah SP (2010) Highly dispersed carbon nanotube reinforced cement based materials. Cem Concr Res 40:1052–1059
26. Peng H, Alemany LB, Margrave JL, Khabashesku VN (2003) Sidewall carboxylic acid functionalization of single-walled carbon nanotubes. J Am Chem Soc 125:15174–15182
27. Xin X, Xu G, Zhao T (2008) Dispersing carbon nanotubes in aqueous solutions by a starlike block copolymer. J Phys Chem C 112:16377–16384
28. Bystrzejewski M, Huczko A, Lange H, Gemming T, Büchner B, Rümmeli MH (2010) Dispersion and diameter separation of multi-wall carbon nanotubes in aqueous solutions. J Colloid Interface Sci 345:138–142
29. Liao YH, Marietta-Tondin O, Liang Z, Zhang C, Wang B (2004) Investigation of the dispersion process of SWNTs/SC-15 epoxy resin nanocomposites. Mater Sci Eng A 385:175–181
30. Li J, Li Z, Tong J, Xiaa C, Li F (2015) Nitrogen-doped ordered mesoporous carbon sphere with short channel as an efficient metal-free catalyst for oxygen reduction reaction. RSC Adv 5:70010–70016
31. Susanto A, Koleva DA, Copuroglu O, van Beek K, van Breugel K (2013) Mechanical, electrical, and microstructural properties of cement-based materials in condition of stray current flow. J Adv Concr Technol 11:119–134
32. Stern M, Geary AL (1957) A theoretical analysis of the shape of polarization resistance curves. J Electrochem Soc 104:56–63
33. Andrade C, Alonso C (1996) Corrosion rate monitoring in the laboratory and on-site. Constr Build Mater 10:315–328
34. Zornoza E, Paya J, Garces P (2008) Chloride-induced corrosion of steel embedded in mortars containing fly ash and spent cracking catalyst. Corros Sci 50:1567–1575
35. Basak R, Bandynopadhyay R (2013) Encapsulation of hydrophobic drugs in pluronic F127 micelles: effect of drug hydrophobicity, solution temperature, and pH. Langmuir 29:4350–4356
36. Eliseeva OV, Besseling NAM, Koopal LK, Cohen Stuart MA (2005) Influence of NaCl on the behavior of PEO-PPO-PEO triblock copolymers in solutions, at interface, and in asymmetric liquid films. Langmuir 21:4954–4963

37. Eliseeva OV, Fokkink RG, Besseling NAM, Koopal LK, Cohen Sturat MA (2006) Thinning of wetting films formed from aqueous solutions of non-ionic surfactant. J Colloid Interface Sci 301:210–216
38. Nita LE, Chiriac AP, Bercea M (2012) Effect of pH and temperatura upon self-assembling process between ploy(aspartic acid) and pluronic F127. Colloid Surf B Biointer 119:6539–6545
39. Alexander S, de Vos WM, Castle TC, Cosgrove T, Prescott SW (2014) Growth and shrinkage of pluronic micelles by uptake and release of flurbiporfen: variation of pH. Langmuir 28:47–57
40. Gong K, Du F, Xia ZH, Durstock M, Dai LM (2009) Nitrogen-doped carbon nanotube arrays with high Electrocatalytic activity for oxygen reduction. Science 323:760–764
41. Deng DH, Pan XL, Yu LA, Cui Y, Jiang YP, Qi J, Li WX, Fu QA, Ma XC, Xue QK, Sun GQ, Bao XH (2011) Toward N-doped graphene via solvothermal synthesis. Chem Mater 23:1188–1193
42. Wan K, Yu ZP, Li XH, Liu MY, Yang G, Piao JH, Liang ZX (2015) pH effect on electrochemistry of nitrogen-doped carbon catalyst for oxygen reduction reaction. Catalysis 5:4325–4432
43. Koleva DA, Denkova AG, Boshkov N, van Breguel K (2013) Electrochemical performance of steel in cement extract and bulk matric properties of cement past in presence of pluronic 123 micelles. J Mater Sci 48:2490–2503
44. Hu J, Koleva DA, de Wit JHW, Kolev H, van Breugel K (2011) Corrosion performance of carbon steel in simulated pore solutions in presence of micelles. J Electrochem Soc 158:C76–C87
45. Diaz B, Joiret S, Keddam M, Novoa XR, Perez MC, Takenouti H (2004) Passivity of iron in mud's water solution. Electrochim Acta 49:3039–3048
46. Joiret S, Keddam M, Novoa XR, Perez MC, Rangel C, Takenouti H (2002) Use of EIS, ring-disk electrode, EQCM and Raman spectroscopy to study the film of oxides formed on iron in 1M NaOH. Cement Concr Comp 24:7–15
47. Freire L, Nóvoa XR, Montemor MF, Carmezim MJ (2009) Study of passive films formed on mild steel in alkaline media by the application of anodic potentials. Mater Chem Phys 114:962–972
48. Alonso C, Andrade C, Izquierdo M, Novoa XR, Perez MC (1998) Effect of protective oxide scales in macrogalvanic behavior of concrete reinforcement. Corros Sci 40:1379–1389
49. Antony H, Legrand L, Marechal L, Perrin S, Dillmann P, Chausse A (2005) Study of lepidor-crocite γ-FeOOH electrochemical reduction in natural and slightly alkaline solutions at 25 °C. Electrochem Acta 51:745–753
50. Freire L, Carmezima MJ, Ferreiraa MGS, Montemora MF (2010) The passive behaviour of AISI 316 in alkaline media and the effect of pH: a combined electrochemical and analytical study. Electrochim Acta 55:6174–6181
51. Koleva DA, Boshkov N, van Breugel K, de Wit JHW (2011) Steel corrosion resistance in model solutions, containing waste materials. Electrochim Acta 58:628–646
52. Andrade C, Merino P, Nóvoa XR, Pérez MC, Soler L (1995) Passivation of reinforcing steel in concrete. Mater Sci Forum 192-194:891–898
53. Mahmoud H, Sanchez M, Alonso MC (2015) Ageing of the spontaneous passive state of 2304 duplex stainless steels in high-alkaline conditions with the presence of chloride. J Solid State Electrochem 19:2961–2972
54. Andrade C, Keddam M, Novoa XR, Perez MC, Rangel CM, Takenouti H (2001) Electrochemical behavior of steel rebars in concrete: influence of environmental factors and cement chemistry. Electrochim Acta 46:3905–3912
55. Abreu CM, Covelo A, Daiz B, Freire L, Novoa XR, Perez MC (2007) Electrochemical behavior of iron in chlorinated alkaline media; the effect of slurries form granite processing. J Braz Chem Soc 18:1156–1163
56. Flis J, Oranowska H, Smialowska ZS (1990) An ellipsometric study of surface films grown on iron and iron carbon alloys in 0.05 M KOH. Corros Sci 30:1085–1099

57. Li WS, Luo JL (1999) Electronic properties and pitting susceptibility of passive films on ferrite and pearlite in chloride-containing solution. Proc Electrochem Soc 27:161–170
58. Macdonald DD (1992) The point defect model for passive state. J Electrochem Soc 139:3434–3449
59. Castro EB (1994) Analysis of the impedance response of passive iron. Electrochim Acta 39:2117–2123
60. Liu L, Xu J, Xie ZH, Munroe P (2013) The roles of passive layers in regulating the electrochemical behaviour of Ti_5Si_3-based nanocomposite films. J Mater Chem A 1:2064–2078
61. Xia Y, Cao F, Liu W, Chang L, Zhang H (2013) The formation of passive films of carbon steels in borate buffer and their degradation behavior in NaCl solution by SECM. Int J Electrochem Sci 8:3057–3073

Chapter 7
Activated Hybrid Cementitious System Using Portland Cement and Fly Ash with Na$_2$SO$_4$

Diego F. Velandia, Cyril J. Lynsdale, John L. Provis, Fernando Ramirez, and Ana C. Gomez

Abstract A number of alternatives have been explored by the cement industry in recent years to reduce CO$_2$ emissions. One of the alternatives, the subject of this chapter, is an intermediate system between a high volume fly ash concrete and a geopolymer concrete. This concrete includes a hybrid system of 50% OPC–50% fly ash, and an activator. In this study, the long-term durability was studied for laboratory and outdoor cured concretes. It was found that chloride diffusion coefficient was reduced significantly at 90 days and beyond for the activated system compared to control samples (100% OPC and 80% OPC–20% fly ash) of the same water to cementitious content ratio (W/CM). This behavior was exhibited by samples cured under laboratory controlled curing conditions (100% RH and 23 °C). On the other hand, outdoor curing increased concrete permeability for all concretes. Long-term carbonation was also explored and samples under outdoor curing had a significant carbonation depth. Alkali silica reaction problems were mitigated with this activated hybrid system. In order to improve the carbonation resistance of this concrete, a reduction in W/CM seems necessary. Based on these results, activated high volume fly ash systems provide a low CO$_2$ concrete alternative; however, more studies are needed for establishing specifications and service life prediction models.

D.F. Velandia (✉)
Department of Civil and Structural Engineering, University of Sheffield,
Sir Frederick Mappin Building, Mappin Street, Sheffield S1 3JD, UK

Research and Development, Argos, Carrera 62 No. 19 – 04, Bogotá, Colombia
e-mail: dvelandia@argos.com.co

C.J. Lynsdale
Department of Civil and Structural Engineering, University of Sheffield,
Sir Frederick Mappin Building, Mappin Street, Sheffield S1 3JD, UK

J.L. Provis
Department of Materials Science and Engineering, University of Sheffield,
Sir Robert Hadfield Building, Mappin Street, Sheffield S1 3JD, UK

F. Ramirez
Department of Civil and Environmental Engineering, Universidad de los Andes,
Carrera 1 este No. 19A-40, Edificio Mario Laserna, Oficina ML 632, Bogotá, Colombia

A.C. Gomez
Research and Development, Argos, Carrera 62 No. 19 – 04, Bogotá, Colombia

© Springer International Publishing AG 2017
L.E. Rendon Diaz Miron, D.A. Koleva (eds.), *Concrete Durability*,
DOI 10.1007/978-3-319-55463-1_7

Keywords Durability • Fly ash • Na$_2$SO$_4$

7.1 Introduction

Deterioration of a concrete structure depends on the permeability of its concrete. Gases, ions and liquids penetrate the concrete, reacting with its constituents and affecting the structure. As a result of physical and chemical interactions, concrete may end in cracks, expansion and matrix deterioration [1–3]. Van Deventer et al. [4] also illustrate how permeability is the main parameter to consider for predicting concrete durability. Pores and cracks present in the matrix (aggregates, cement paste–aggregate interface and cement paste) affect different concrete mechanical properties such as strength and modulus of elasticity [2].

Ingress of different liquids, gases and ions in concrete and their movement inside the matrix are basically due to absorption, permeability and diffusion [3]; these processes depend on external pressure, chemical concentration gradient in solution and the moisture state of the concrete.

This study focuses on the above durability parameters in evaluating the performance of mixes with hybrid systems, including 50% FA and 1% sodium sulfate as activator, and different water to cementitious materials ratios. Table 7.1 presents the scope of the study and coding system used to identify mix and sample parameters.

7.2 Durability Evaluation

In general terms, the curing process had an effect on water permeability, especially in mixes with high volumes of fly ash. Samples cured outdoors showed an increase in water permeability. Laboratory curing guaranteed the water for the hydration of cement and the formation of hydration products. Outdoor conditions such as relative humidity and temperature were not as favorable for hydration product formation. As seen in Fig. 7.1, in most of the cases, specimens with activator presented lower water permeability than control samples at 180 days. The effect of water to cementitious material ratio on water permeability was significant for mixes with fly ash. This was significant for mixes with lab curing; mixes under outdoor conditions did not have a pattern in their behavior.

Laboratory cured fly ash mixes benefited from the presence of activator where it can be seen in Fig. 7.2 that diffusion values are significantly lower in comparison with nonactivated mixes and generally lower than for the control cement mixes with the marked effect seen by 90 days in both cases. Although the reduction in diffusion values in nonactivated mixes lagged behind other systems, with time there was a significant improvement.

Although carbonation depth was lower for mixes with 50% fly ash and activator than those without it, there was no significant difference between them as seen in Fig. 7.3. Carbonation depths for mixes with 20% fly ash did not change with time

Table 7.1 Mix ID (**a**) Code order and Description, (**b**) code variables

(a)	
Mix ID (1/2/3/4/5)	
Letters and numbers order	Description
1	W/CM
2	Cementitious material type
3	Cementitious material percentage
4	Curing type
5	Activator

(b)	
1. W/CM	
0.675	
0.557	
0.483	
0.426	
2. Cementitious material type	
CE	Portland cement
TP	Termopaipa FA
3. Cementitious material percentage	
0	0%
20	20%
50	50%
100	100%
4. Curing type	
L	Lab curing
O	Outdoor curing
5. Activator	
A	Sodium sulfate

significantly as for 50% fly ash mixes. The low portlandite content in mixes with 50% of fly ash made specimens to carbonate faster than 100% cement concretes.

Based on Fig. 7.4, mixes with 50% of fly ash had an expansion lower than 0.1% at 16 days. Even after 30 days, expansion was lower than 0.1% for mixes including 50% of fly ash. According to Fig. 7.4, the specimen with activator had a similar behavior to the one with 50% of fly ash only.

7.3 Conclusions

Mixes with fly ash and sodium sulfate were either comparable or superior to the control cement concrete of the same W/CM in terms of water permeability and chloride diffusion coefficient, when water cured. Outdoor curing adversely affected

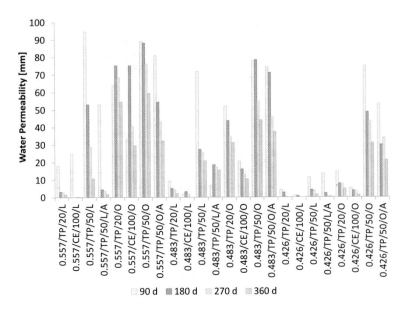

Fig. 7.1 Water permeability

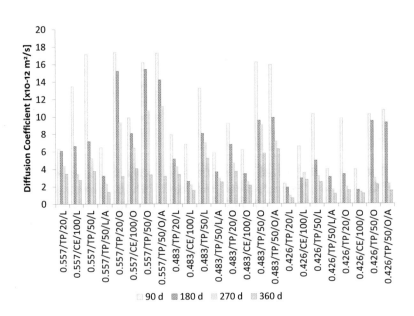

Fig. 7.2 Chloride diffusion coefficient

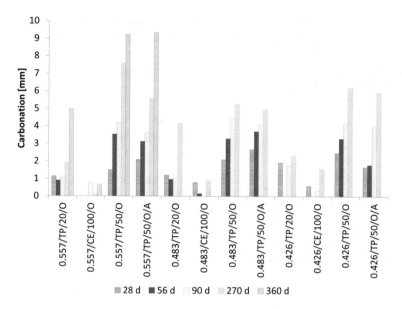

Fig. 7.3 Carbonation depths

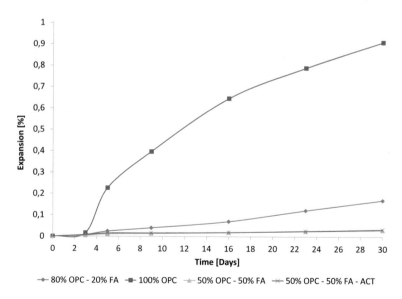

Fig. 7.4 Alkali silica reaction evaluation

the performance of the fly ash concrete. Values of carbonation were not favorable for mixes with sodium sulfate; however, the performance of fly ash mixes could be improved, to match those of the control, by reducing the water to cementitious content ratio. Specimen expansion due to alkali silica reaction was lower for mixes with sodium sulfate.

References

1. Mehta PK, Gerwick BC (1982) Physical causes of concrete deterioration. Concr Int 4:45–51
2. Basheer L, Kropp J, Cleland D (2001) Assessment of the durability of concrete from its permeation properties: a review. Constr Build Mater 15:93–103
3. Long AE, Henderson GD, Montgomery FR (2001) Why assess the properties of near-surface concrete? Constr Build Mater 15:65–79
4. Van Deventer JSJ, Provis JL, Duxson P (2012) Technical and commercial progress in the adoption of geopolymer cement. Miner Eng 29:89–104

Chapter 8
Optimum Green Concrete Using Different High Volume Fly Ash Activated Systems

Diego F. Velandia, Cyril J. Lynsdale, Fernando Ramirez, John L. Provis, German Hermida, and Ana C. Gomez

Abstract Environmental issues related to CO_2 emissions have become a key focus for many different industries, including the cement and concrete industry. An environmentally optimized 'green' concrete can provide a much needed alternative to conventional concrete to reduce the carbon footprint of the construction industry. This can be achieved through high Portland cement replacement by fly ash and with the inclusion of activators to enhance the rate of development of strength and other properties. This study evaluates different fly ashes and different activators (Na_2SO_4, lime and quicklime) that are added to enhance the reaction of the fly ash to achieve a comparable performance to that of standard Portland cement in mixes of much lower CO_2 emissions. TGA, XRD and SEM are used to determine the development of hydration products and the consumption of portlandite by the fly ash. It is found that the amorphous content of the fly ash is an important parameter influencing compressive strength evolution. Based on the results, Na_2SO_4 as an activator, and a fly ash with high reactive SiO_2 and Al_2O_3 contents and low Fe_2O_3 are found to

D.F. Velandia (✉)
Department of Civil and Structural Engineering, University of Sheffield,
Sir Frederick Mappin Building, Mappin Street, Sheffield S1 3JD, UK

Research and Development, Argos, Carrera 62 No. 19 – 04, Bogotá, Colombia
e-mail: dvelandia@argos.com.co

C.J. Lynsdale
Department of Civil and Structural Engineering, University of Sheffield,
Sir Frederick Mappin Building, Mappin Street, Sheffield S1 3JD, UK

F. Ramirez
Department of Civil and Environmental Engineering, Universidad de los Andes,
Carrera 1 este No. 19A-40, Edificio Mario Laserna, Oficina ML 632, Bogotá, Colombia

J.L. Provis
Department of Materials Science and Engineering, University of Sheffield,
Sir Robert Hadfield Building, Mappin Street, Sheffield S1 3JD, UK

G. Hermida
Sika Technical Support, km 20,5 Autopista Norte, Tocancipa, Colombia

A.C. Gomez
Research and Development, Argos, Carrera 62 No. 19 – 04, Bogotá, Colombia

© Springer International Publishing AG 2017
L.E. Rendon Diaz Miron, D.A. Koleva (eds.), *Concrete Durability*,
DOI 10.1007/978-3-319-55463-1_8

provide the best options for producing a high volume fly ash matrix with the potential to show comparable behavior to a Portland cement control mix.

Keywords Green concrete • Activated fly ash

8.1 Introduction

In recent years, different possible solutions to reduce the carbon footprint of concrete have been studied by cement and concrete researchers [1–13]. One of the solutions put forward is the inclusion of supplementary cementitious materials in high percentages for concrete production [2, 3]. To advance this concept, all important parameters need to be optimized to develop an optimum green concrete (UOGC) as a low CO_2 concrete alternative for the construction industry. The UOGC, investigated in this work, is based on a high volume fly ash concrete (HVFA), with added alkaline activators. Although this technology has been explored before [4], there are still many unanswered questions [5] which relate to why it has never yet been produced in sufficient volumes to compete with regular Portland cement concrete. Lack of detailed technical information, standards, and the evident need to further research its fresh and hardened properties and durability [6] are some of the reasons motivating further work aimed to explore and develop answers regarding its real viability.

In this study, which is primarily of a scoping nature, mortars and pastes will be evaluated including different activators (sodium sulfate, lime and quicklime) and different high loss on ignition (LOI) fly ashes. Based on compressive strength, X-ray fluorescence (XRF), X-ray diffraction (XRD), thermogravimetry and scanning electron microscopy (SEM) analysis, the preferred activator and its optimum dosage will be determined.

8.2 Experimental Details

Fly ashes (FA) included in this study are referenced as TP FA, FB FA, TG FA and TA FA; the main difference between them is that TP FA, FB FA and TG FA include a high LOI content, while TA FA has a low content. Initially each fly ash was sieved in order to see the effect of the fineness increment and the variations of their compositions. After that, mortar and pastes were produced including different activator dosages with each fly ash, a Type III cement (ASTM C 150) and considering as constant the water to cementitious material ratio.

A PANalytical Axios sequential wavelength dispersive XRF (WDXRF) was used to obtain the chemical composition of each different fly ash. Furthermore, a PANalytical (X'PERT-PRO MPD) system was used for the fly ash mineralogical XRD evaluation. In order to determine the amorphous content, the Rietveld method was followed, using rutile as the internal standard. For mortar mixes, most of the standard ASTM C 109 procedure was followed, with some additional considerations about the mixing of materials; all of the activators were always added to

Table 8.1 Mix IDs (**a**) order and description (**b**) code for each variable

(a)

Mix ID (1/2/3/4/5/6)	
Letters and numbers order	*Description*
1	Cementitious material name
2	Fly ash size
3	Fly ash percentage
4	Activator
5	Dosage
6	Age

(b)

1. Cementitious material name	
CE	Cement
TP	FA 1
FB	FA 2
TG	FA 3
TA	FA 4
2. Size	
OS	Original size – Fly ash
10	10 μm (D50 – Cement)
3. Fly percentage	
0	0%
20	20%
50	50%
100	100%
4. Activators	
A	Sodium sulfate
Q	Quicklime
L	Lime

water and mixed before adding the cementitious materials. As expected in this process, quicklime was the only material which increased the temperature significantly. Portlandite was quantified using thermogravimetry analysis and considering the mass change between 450 and 550 °C.

Table 8.1 presents how different codes describing mix design parameters are organized in the mix IDs throughout this study. It is necessary to use the codes and mix IDs due to the number of parameters evaluated in this work.

8.3 Results and Discussion

Paya et al. mentioned how the reactivity of fly ash increased by improving its fineness [7]. In this study and according to the results shown in Fig. 8.1, the amorphous content of fly ash has a strong influence on the compressive strength. By improving fly ash fineness its composition was changed; Table 8.2 summarizes the effect of the sieving process on the main parameters of each ash. The amorphous, silica and LOI contents were different for each fineness in each case. Figure 8.1 shows that the compressive strength of mortar samples with 20% of fly ash was improved when the amorphous content increased. According to Duran, the compressive strength decreases by increasing the LOI content [3], but in the present study there were some unexpected tendencies where even when the particle size and the LOI content decreased, the compressive strength decreased and it was due because the amorphous content was lower. At some point, fly ashes would not need a mechanical treatment based on the initial amorphous content.

Considering the interaction with activators at different dosages, TP FA had the best performance with sodium sulfate at a dosage of 1%; according to Fig. 8.2a, compressive strength increased about 40% compared to the sample without activa-

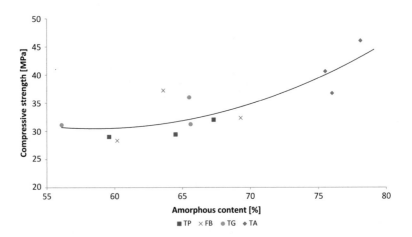

Fig. 8.1 Influence of the amorphous content of the fly ashes on the 28-day compressive strength of samples with 20% cement replacement by fly ash

Table 8.2 Changes in fly ash properties with sieving

Fly ash	Sieve-treatment	Main parameters				
		LOI	Fe_2O_3	CaO	SiO_2	Amorphous
TPFA	–	10.74	4.92	3.27	56.67	64.5
	<74 μm	8.67	5.90	0.57	59.502	67.3
	<45 μm	5.07	5.25	1.43	62.307	59.6
FBFA	–	12.00	4.39	5.99	43.83	69.3
	<74 μm		3.82	3.20	44.96	60.2
	<45 μm	5.78	4.76	6.94	45.445	63.6
TGFA	–	8.74	9.77	3.64	55.14	65.6
	<74 μm	1.54	11.15	2.57	63.119	56.1
	<45 μm	1.94	10.46	4.37	56.892	65.5
TAFA	–	8.74	9.77	3.64	58.58	65.6
	<74 μm	1.54	11.15	2.57	57.92	56.1
	<45 μm	1.94	10.46	4.37	56.59	65.5

tor. Qian et al. evaluated the effectiveness of this activator with HVFA mixes and found that Na_2SO_4 reacts directly over the $Ca(OH)_2$, increasing the alkalinity and accelerating fly ash dissolution; SO_4 increases the formation of ettringite, affecting the density of the mortar matrix positively [8]. The XRD results show that sodium sulfate addition led to the formation of more ettringite than the other activators, improving the strength of this mix.

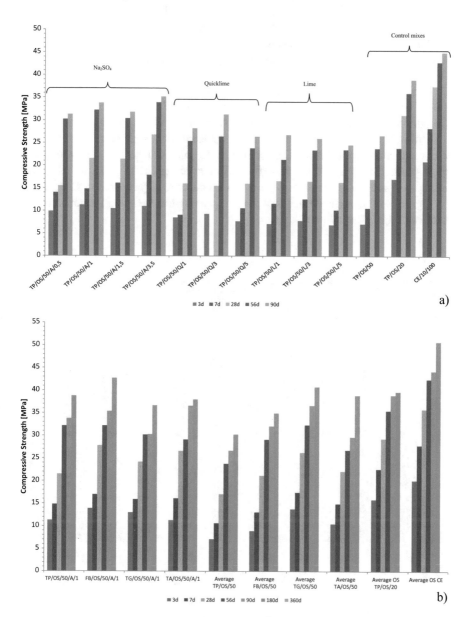

Fig. 8.2 Compressive strength evolutions (**a**) TP FA with sodium sulfate, lime, quicklime and Na_2SO_4, (**b**) fly ashes with Na_2SO_4 at 1%

Sodium sulfate was the activator which presented the best performance using different fly ashes. The influence of fly ashes was also relevant; for instance, FB FA and TP FA were the most reactive for this activator (Fig. 8.2b). The compressive strength performance was the first indicator of their effectiveness. Initially it was expected that TA fly ash would have the best performance in the presence of activators due to its low LOI, but TA did not react as well as FB FA, as presented in Fig. 8.2b; this is possibly due to the high Fe_2O_3 content, which reduced the speed of dissolution of the reactive components of fly ash. Fernandez-Jiménez and Palomo presented some results with high Fe_2O_3 content in fly ash; Fe_2O_3 did not appear in the products of the main reactions [9]. On the other hand, FB fly ash had the lowest Fe_2O_3 content and one of the highest compressive strength values using sodium sulfate (Fig. 8.2b).

Portlandite consumption in mixes with sodium sulfate and 50% of TP FA started after 3 days of age, compared to mixes with FB FA at 7 days. Figure 8.3 shows that the portlandite content of mixes with TG FA and TA FA kept increasing at 28 days.

Ettringite and the amorphous content calculated with XRD were coherent with the compressive strength evolution; at the first days the formation of ettringite helped to improve the compressive strength, and at later ages the formation of C-S-H increased, part of it included in the amorphous content, as presented in Table 8.3.

Figure 8.4 shows SEM images of gypsum, ettringite and C-S-H formation on the surface of TP FA at different ages.

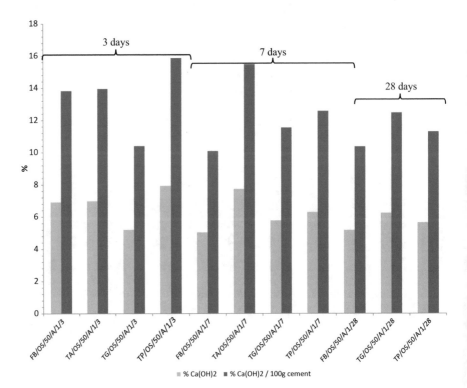

Fig. 8.3 Ca(OH)$_2$ content using TGA: fly ashes with Na$_2$SO$_4$ at 1%

Table 8.3 XRD analysis of fly ashes with Na_2SO_4 at 1% and 3, 7 and 28 days

Mix ID	Quartz low	Tobermorite 9A	Mullite	Portlandite	Ettringite	$C_2S + C_3S$	C_3A	Amorphous content
TP/OS/50/A/1/7	8.3	5.4	7.6	5.7	5.8	3.80	0.2	57.7
TP/OS/50/A/1/3	7.3	7.4	7.8	6.0	5.6	4.30	0.2	58.1
TP/OS/50/A/1/28	5.76	8.08	5.35	2.65	2.17	2.44	0	71.39
TG/OS/50/A/1/7	6.9	5.06	2.69	7.19	2.18	7.71	0.27	67.99
TG/OS/50/A/1/3	4.84	3.69	2.41	5.46	2.52	8.15	0.57	72.37
TG/OS/50/A/1/28	5.53	6.75	2.12	5.18	1.82	3.36	0.03	75.22
TA/OS/50/A/1/7	8.51	8.49	4.57	9.38	1.06	7.85	0.15	60
TA/OS/50/A/1/3	7.99	6.37	4.73	7.9	1.09	10.69	0.49	60.74
FB/OS/50/A/1/7	2.4	6.6	9.1	3.6	4.8	3.87	0.2	65.2
FB/OS/50/A/1/3	2.5	4.8	8.2	3.7	4.3	4.47	0.3	69.2
FB/OS/50/A/1/28	2.58	11.37	7.37	2.61	2.9	3.61	0.07	66.9

Fig. 8.4 SEM images of (**a**) TP/OS/50/A/1/7, (**b**) TP/OS/50/A/1/28

8.4 Conclusions

The addition of sodium sulfate to high volume fly ash-Portland cement binders increases ettringite formation and portlandite consumption; these characteristics were evident in the compressive strength evolution, thermogravimetry and XRD results; on the other hand, quicklime and lime did not present any positive effect in the activation process.

Initially it was evident that a high amorphous content in the fly ash could help to increase the compressive strength in mixes without activators. After considering mixes with activators, the influence of the fly ash Fe_2O_3 was also relevant, as was evident with high Fe_2O_3 fly ashes, where it seems that the speed of dissolution of the fly ash decreased affecting the activation process negatively. These results provide initial steps toward the design and optimization of hybrid high-volume fly ash Portland cement-alkaline cements and mortars, aiming towards the development of Ultra-Optimum Green Concrete for sustainable development in the construction industry.

References

1. Pade C, Guimaraes M (2007). The CO2 uptake of concrete in a 100 year perspective. Cem Concr Res 37:1348
2. Malhotra VM (2002). High-performance, high-volume fly ash concrete. Concr Int 24:30
3. Atiş CD (2005). Strength properties of high volume fly ash roller compacted and workable concrete, and influence of curing condition. Cem Concr Res 35:1112
4. Palomo A, Grutzeck MW, Blanco MT (1999). Alkali-activated fly ashes. A cement for the future Cem Concr Res 29:1323
5. Pacheco-Torgal F, Abdollahnejad Z, Camoes AF, Jamshidi M, Ding Y (2012). Durability of alkali-activated binders: A clear advantage over Portland cement or an unproven issue? Const Building Mat 30:400
6. Shi C, Fernández-Jimenez A, Palomo A (2011). New cements for the 21st century: The pursuit of an alternative to Portland cement. Cem Concr Res 41:750

7. Paya J, Monzó J, Borrachero MV, Peris-Mora E, Gonzalez-Lopez E (1997). Mechanical treatment of fly ashes. Part III: Studies on strength development of ground fly ashes (GFA) – cement mortars. Cem Concr Res 27:1365
8. Qian J, Shi C, Wang Z (2001). Activation of blended cements containing fly ash. Cement and Concrete Research. Cem Concr Res 31:1121
9. Fernández-Jimenez A, Palomo A (2003). Characterization of fly ashes. Potential reactivity as alkaline cements. Fuel 82:2259
10. Criado M, Palomo A, Fernández-Jimenez A (2005). Characterization of fly ashes. Potential reactivity as alkaline cements, Fuel 84:2048
11. Poon C, Kou S, Lam L, Lin Z (2001). Activation of fly ash/cement using calcium sulfate anhydrite (CaSO4). Cem Concr Res 31:873
12. Lee CY, Lee HK, Lee KM (2003). Strength and microstructural of chemically activated fly ash-cement systems. Cem Concr Res 33:425
13. Antiohos S, Papageorgiou A, Papadakis V, Tsimas S (2007). Influence of quicklime addition on the mechanical properties and hydration degree of blended cements containing different fly ashes. Const Building Mat 22:1191

Index

© Springer International Publishing AG 2017
L.E. Rendon Diaz Miron, D.A. Koleva (eds.), *Concrete Durability*,
DOI 10.1007/978-3-319-55463-1

Printed in the United States
By Bookmasters